A History of the Mishnaic Law of Agriculture

BROWN UNIVERSITY
BROWN JUDAIC STUDIES

Edited by

Jacob Neusner
Wendell S. Dietrich, Ernest S. Frerichs,
Horst R. Moehring, Sumner B. Twiss

Number 18

A History of the Mishnaic Law of Agriculture
Tractate Maaser Sheni

by Peter J. Haas

A History of the Mishnaic Law of Agriculture

Tractate Maaser Sheni

by
Peter J. Haas

Scholars Press

Distributed by
SCHOLARS PRESS
101 Salem Street
Chico, CA 95926

A History of the Mishnaic Law of Agriculture
Tractate Maaser Sheni

by
Peter J. Haas

Library of Congress Cataloging in Publication Data

Haas, Peter J
A history of the Mishnaic law of agriculture.

(Brown Judaic studies ; 18)
"In the following study I translate and explain Tractate Maaser sheni...in Mishnah...and Tosefta."
Bibliography: p.
Includes indexes.
1. Mishnah. Ma'aser sheni—Commentaries. 2. Tosefta. Ma'aser sheni—Commentaries. 3. Tithes (Jewish law) I. Mishnah. Ma'aser sheni. English. 1980. II. Tosefta. Ma'aser sheni. English. 1980. III. Title. IV. Series.
BM506.M15H3 296.1'2306 80-25479
ISBN 0-89130-442-8
ISBN 0-89130-443-6 (pbk.)

Printed in the United States of America
1 2 3 4 5
McNaughton & Gunn
Ann Arbor, Michigan 48106

To Lee

CONTENTS

PREFACE

In the following study, I translate and explain Tractate Maaser Sheni ("second tithe") in Mishnah (hereafter M.) and its supplement, Tosefta (T.).[1] Second tithe is an agricultural offering which Israelite farmers must separate from their annual harvest, bring to Jerusalem and eat there. The framers of M. derive their notion of this tithe from Deut. 14:22-27. According to Scripture, the farmer may either carry the produce itself to the holy city or he may instead bring its value in money. He then uses this money in Jerusalem to buy food to be eaten in place of the original produce. M.'s concern is to determine how the farmer is to handle the tithe so that it is preserved until properly eaten. That is, the farmer must take care not to lose, destroy or misuse any portion of produce in the status of second tithe. If he sells the produce, he must be sure to receive in exchange its full value in coin. This is to insure that he will buy the proper amount of food in Jerusalem. The point of M. is that, in either case, the total value of the tithe ultimately must be eaten.

The purpose of my study is to discover what the authorities who formulated the laws meant to say by them. As a first step, I offer a fresh translation of the tractate in both M. and T. In my translation, I pay close attention to the syntax and patterned language of the original Hebrew. I then allow these formal traits to guide my reading and interpretation of the law's substance. Through attention to the law's formal characteristics, as well as to its contents, I claim to adduce the point of law the formulator intended to make. Following this analysis, I step back and identify the general themes or topics which the redactor discusses in the tractate as a whole. This I do by noting the sequence of units established by the redactor and adducing the logic by which these have been arranged. The final stage in my study is to describe the central issue, or problematic, with which the tractate deals. Although these are the last two steps in my study, I present the results in the Introduction so that the reader is oriented to the law as a whole before turning to the details of each pericope.

Through my study of the tractate, I hope to contribute to the larger goal of describing the world view of early rabbinic Judaism. My contention is that the creation of Mishnah as a whole at the end of the second century represents an attempt on the part of its authorities to come to terms both with the destruction of the cult in 70 and with the end of all hope of reestablishing it when the Bar Kokhba revolt failed. My interest is in the extent to which the larger program of Mishnah's authorities is reflected in their discussion of second tithe. Second tithe is a particularly appropriate topic for this purpose, since it concerns produce which the farmer grows and controls, and yet of which he must somehow dispose in accordance with the laws of holiness. By investigating how M. wants such obligations fulfilled, we gain insight into the kind of society M.'s authorities envisioned for Israel.

It is with profound gratitude that I acknowledge the support and help of others who have made this work possible. Foremost among these is my teacher Jacob Neusner. He patiently guided me through this work page by page as it was being prepared. His insightful comments and suggestions prodded me to surpass myself. I owe him an immense debt for all that he has taught me over the last three years. In addition I would like to express my gratitude to the members of the Religious Studies Department at Brown University, each of whom has contributed to my intellectual growth. I mention in particular Prof. Wendell S. Dietrich, Ernest S. Frerichs (Dean of the Graduate School), Horst R. Moehring, John P. Reeder, Jr., Sumner B. Twiss, Jr., J. Giles Milhaven, R. S. Sarason (now of Hebrew Union College-Jewish Institute of Religion), David Goodblatt (of Haifa University) and Dean McBride (of Northwestern University). In addition to their other duties, Dean Frerichs and David Goodblatt agreed to serve as readers for my dissertation. I am thankful for their time and energy. Lois Atwood and Evelyn Palombo, administrative assistants in the Department of Religious Studies did much to make my stay at Brown a pleasant experience.

The members of Prof. Neusner's graduate seminar patiently sat through successive rewritings of this manuscript. Leonard Gordon, Abraham Havivi, Martin Jaffee (University of Virginia), Irving Mandelbaum (University of Texas) and Alan Peck each offered comments and suggestions which contributed immeasurably to this work. I would like to thank in particular Alan Peck, who generously helped me in preparing earlier drafts of this study, and Martin Jaffee, who reviewed the entire manuscript and suggested many ways it could be improved. Their insight and friendship has been invaluable. They certainly bear no responsibility for any deficiency in this work.

Above all, this dissertation is a tribute to my wife, Lee. Despite the demands of raising our family, she found time and energy to help me with typing involved in producing this study. Words cannot express the gratitude I feel for the support, moral and otherwise, that she gave me during these years of graduate study. In dedicating this work to her, I only give to her a small part of what she has given to me.

Peter Jerome Haas
20 June, 1980
6 Tammuz, 5740

ABBREVIATIONS AND BIBLIOGRAPHY

Albeck = Hanoch Albeck, _Šišah Sidré Misnah, Seder Zera^cim_ (Jerusalem and Tel Aviv, 1957)

Arak. = ^cArakin

Aruch = Alexander Kohut, ed., _Aruch Completum_, 8 vols. (Vienna 1878-92; second edition, 1926), and _Supplementary Volume_, ed., Samuel Krauss _et al_. (Vienna, 1937)

A.Z. = ^cAbodah Zarah

b. = _Babli_, Babylonian Talmud; _ben_, "son of," as in "Simeon b. Gamaliel"

Bacher = Wilhelm Bacher, _Die Agada der Tannaiten_ (Strassling, 1903)

B.B. = Baba' Batra'

Beer-Holtzmann = G. Beer and O. Holtzmann, eds., _Die Mischna: Text, Übersetzung und ausführliche Erklärung_, I. 7-8 _Seder Zera^cim: Maaserot/Maaser Sheni_ by Wolfgang Bunte (Berlin, 1962)

Bek. = Bekhorot

Bert = Obadiah b. Abraham of Bertinoro, d.c. 1500, Commentary to Mishnah, in Romm edition of Mishnah

Bes. = Beṣah

Bik. = Bikkurim

Blackman = _Mishnayoth, I. Order Zeraim_. Pointed Hebrew text, English translation, introductions, notes ... by Philip Blackman (London, 1955; second edition, New York, 1964)

B.M.	=	Baba' Meṣi^{c}a'
B.Q.	=	Baba' Qamma'
B.R.	=	Julius Theodor and Chanoch Albeck, eds. *Midrasch Bereschit Rabbah* (Berlin, 1903-1936)
Bunte	=	See Beer-Holtzmann
Celsus	=	Aulus Cornelius Celsus, *De Medicina*, W. G. Spencer, trans. (Cambridge, MA, 1935-1938)
Columella	=	Columella, *De Re Rustica*, trans. H. B. Ash, et al. (London, 1941-1955)
D	=	Tosefta, *editio princeps* (Venice, 1521)
Dalman, *Arbeit*	=	Gustav Dalman, *Arbeit und Sitte in Palästina*, 8 vols. (Gutersloh, 1928-1942)
Danby	=	*The Mishnah*, trans. Herbert Danby (London, 1933)
Dem.	=	Dem'ai
Deut.	=	Deuteronomy
Dorland	=	W. A. N. Dorland, ed., *Dorland's Illustrated Medical Dictionary* (Philadelphia, 1957)
E	=	Tosefta, MS. Erfurt., See Lieberman, pp. 8-11
Ed.	=	^{c}Eduyot
EJ	=	*Encyclopaedia Judaica* (Jerusalem, 1972)
Er.	=	^{c}Erubin
Ex.	=	Exodus
Feliks	=	Yehuda Feliks, *Ha-Ḥaqla'ut be'ereṣ yiśra'el bitqupat ha-mišnah veha-talmud* (Tel Aviv, 1963)

Gereboff	=	Joel Gereboff, Rabbi Tarfon: The Tradition, the Man, and Early Rabbinic Judaism (Missoula, 1979)
Git.	=	Giṭṭin
GRA	=	Elijah b. Solomon Zalman ("HaGa'on Rabbi 'Eliyahu," or "Vilna Gaon," Lithuania, 1720-1797), Mishnah commentary in Romm edition of Mishnah (Vilna, 1908); Tosefta emendations in Romm edition of Babylonian Talmud (Vilna, 1886)
Green	=	W. S. Green, ed., Approaches to Ancient Judaism, Vol. I (Missoula, 1978), Vol. II (Missoula, 1980)
Guthrie	=	H. H. Guthrie, Jr., s.v., "Tithes," in H. Buttrick, ed., The Interpreter's Dictionary of the Bible (Nashville, 1962), 4:654-655.
Ḥag	=	Ḥagigah
HD	=	Hasdé David, David Samuel b. Jacob Pardo, ed. [Tosefta commentary], I. Seder Zera^{c}im (Livorno, 1776)
Hul.	=	Ḥulin
Jaffee	=	Martin Jaffee, A Study of Tractate Maaserot, doctoral dissertation, Brown University
Jastrow	=	Marcus Jastrow, A Dictionary of the Targumim, the Talmud Babli and Yerushalmi and the Midrashic Literature, 2 vols. (New York, 1895-1903)
JE	=	The Jewish Encyclopedia, ed. Isidore Singer (New York, 1901)
Kasovsky	=	Chayim Yehoshua Kasovsky, Thesaurus Mishnae: Concordantiae verborum etc., 4 vols., rev. ed. (Tel Aviv, 1967)

Kasovsky, *Tosefta* = Chayim Yehoshua Kasovsky, *Thesaurus Thosephthae: Concordantiae verborum etc.*, 6 vols. (Jerusalem, 1932-1961)

Kel. = Kelim

Ket. = Ketubot

Kindler = Arie Kindler and Ernst Klimowsky, *The Function and Pattern of the Jewish Coins and the City-coins of Palestine and Phoenicia* (Jerusalem, 1968)

Klimowsky = Ernst Klimowsky, *On Ancient Palestinian Coins, their Symbolism and Metrology* (Tel Aviv, 1974)

KM = Joseph b. Ephraim Karo, *Kesep Mišneh*, commentary to Maimonides' *Mishneh Torah*, in standard editions of the latter

Krauss = Samuel Krauss, *Talmudische Archäologie*, 3 vols. (Leipzig, 1910-1912)

Lev. = Leviticus

Lieberman = Saul Lieberman, comm., *The Tosefta According to Codex Vienna, with Variants from Codex Erfurt, Genizah MSS. and Editio Princeps*, I. *The Order of Zera^{c}im* (New York, 1955)

Lieberman, *Hellenism* = Saul Lieberman, *Hellenism in Jewish Palestine* (New York, 1950)

Lieberman, *TK* = Saul Lieberman, *Tosefta Ki-fshutah: A Comprehensive Commentary on the Tosefta*, I. *Order Zera^{c}im*, 2 vols. (New York, 1955)

Loew, *AP* = Immanuel Loew, *Aramäische Pflanzenname* (Leipzig, 1881)

Loew, *Flora* = Immanuel Loew, *Die Flora der Juden*, 4 vols. (Vienna and Leipzig, 1926)

Ma.	=	Ma^{c}aserot
Maim.	=	Maimonides (Moses b. Maimon 1135-1204), Kitab es-Siraj [Mishnah commentary] in Romm edition of Mishnah
Mak.	=	Makkot
MB	=	Minḥat Bikkurim, comm. Samuel Avigdor b. Abraham Karlin, in Romm ed. of Babylonian Talmud
Me	=	Me^{c}ilah
Meg.	=	Megillah
Mekh.	=	Mechilta d'Rabbi Ismael, ed. H. S. Horovitz and I. Rabin (Jerusalem, 1970)
Men.	=	Menaḥot
M.Q.	=	Mo^{c}ed Qatan
MR	=	Mišnah Ri'šonah, comm. Ephraim Isaac of Premysla (Poland) in Romm ed. of Mishnah
M.S.	=	Ma^{c}aser Šeni
MS	=	Meleket Šelomoh, comm. Solomon b. Joshua Adeni in Romm ed. of Mishnah
Ned.	=	Nedarim
Neg.	=	Nega^{c}im
Neubaur	=	Adolf Neubaur, La Géographie du Talmud (Paris, 1868)
Neusner, Eliezer	=	Jacob Neusner, Eliezer ben Hyrcanus: The Tradition and the Man, 2 vols. (Leiden, 1973)
Neusner, Holy Things	=	Jacob Neusner, A History of the Mishnaic Law of Holy Things, 6 vols. (Leiden, 1978-1979)

Neusner, Pharisees	=	Jacob Neusner, The Rabbinic Traditions about the Pharisees before 70, 3 vols. (Leiden, 1971)
Neusner, Purities	=	Jacob Neusner, A History of the Mishnaic Law of Purities, 22 vols. (Leiden, 1974-1978)
Neusner, Tosefta	=	Jacob Neusner, The Tosefta Translated from the Hebrew, vol. III, "The Order of Women"; vol. IV, "The Order of Holy Things"; vol. V, "The Order of Purities"; (New York, 1977-1979)
NH	=	Caius Secundus Pliny, Natural History (Cambridge, Mass., 1938-1942)
Nid.	=	Niddah
Or.	=	ᶜOrlah
Pe.	=	Pe'ah
Peck	=	Alan Peck, A Study of Tractate Terumot, doctoral dissertation, Brown University
Pes.	=	Pesaḥim
PM	=	Moses b. Simeon Margoliot (d. 1781), Penei Moshe, commentary to Palestinian Talmud; in Zhitomir ed. of latter
Porton	=	Gary Porton, The Traditions of Rabbi Ishmael (Leiden, 1976-)
Press	=	Isaiah Press, 'Ereṣ Yiśra'el: Enṣyqlopediah Topograpit historit, 4 vols. (Jerusalem, 1951-1955)
Preuss	=	Julius Preuss, Biblisch-talmudische Medizin (Berlin, 1911)
Primus	=	Charles Primus, Aqiva's Contribution to the Law of Zeraᶜim (Leiden, 1977)
Qid.	=	Qiddušin

Rabad = Abraham b. David of Posquières (Provence, c. 1120-1198), glosses to Maimonides' *Mishneh Torah* in standard editions of the latter

RGG[3] = *Die Religion in Geschichte und Gegenwart* Handwörterbuch für Theologie und Religionswissenschaft, 3rd ed. (Tuebingen, 1957)

R.H. = Roš Hašanah

Sacks-Hutner = Nissan Sacks, ed., *The Mishnah with Variant Readings, Order Zera^c^im*, 2 vols. Edited at Institute for the Complete Israeli Talmud, Joshua Hutner, director (Jerusalem, 1972-1975)

San. = Sanhedrin

Sarason, *Demai* = R. S. Sarason, *A History of the Mishnaic Law of Agriculture: A Study of Tractate Demai* (Leiden, 1979)

Sarason, Mishnah and Scripture = R. S. Sarason, "Mishnah and Scripture: Preliminary Observations on the Law of Tithing in *Seder Zera^c^im*," in W. S. Green, ed., *Approaches to Ancient Judaism* II (Missoula, 1980)

SCH = *Standard Cyclopedia of Horticulture* 6 vols. (London, 1915)

Sens = Samson b. Abraham of Sens (France, late twelfth-early thirteenth centuries) Mishnah commentary in Romm ed. of Babylonian Talmud

Shab. = Šabbat

Sheb. = Šebi^c^it

Sheq. = Šeqalim

Sifra = *Sifra debe Rab, hu' Sefer Torat Kohanim*, ed. I. H. Weiss (Vienna, 1862)

Sif. D.	=	Louis Finkelstein, ed., *Siphre on Deuteronomy* (Berlin, 1939)
Sif. Nu.	=	*Siphre debe Rab fasciculus primus: Siphre ad Numeros adjecto Siphre Zutta*, ed. H. S. Horovitz (Leipzig, 1917)
Sirillo	=	Solomon b. Joseph Sirillo (d. ca. 1558), commentary on Palestinian Talmud
Soncino	=	*The Babylonian Talmud: Seder Zera^{c}im*, II. *Maaser Sheni*, ed. Moses Hirsch Segal (London, 1948)
Sot.	=	Soṭah
Tem.	=	Temurot
Ter.	=	Terumot
Toh.	=	Tohorot
TYB	=	Baruch Isaac b. Israel Lipschütz (1812-1877), *Tip'eret Yisrael Bo^{c}az*, super-commentary to TYY in Romm ed. of Mishnah
T.Y.	=	Ṭebul Yom
TYT	=	Yom Tob Lippmann Heller (1579-1654), *Tosepot Yom Tob*, Mishnah commentary in Romm ed. of Mishnah
TYY	=	Israel b. Gedaliah Lipschütz (1782-1860), *Tip'eret Yiśrael Yakin*, Mishnah commentary in Romm ed. of Mishnah
Uq.	=	^{c}Uqṣin
White	=	E. B. White, *Roman Farming* (Ithaca, N.Y., 1970)
y.	=	*Yerušalmi*, Palestinian Talmud
Yad.	=	Maimonides, *Mishné Torah*, standard ed., 6 vols.

Yeb.	=	Yebamot
Zahavy	=	Tsvee Zahavy, The Traditions of Eleazar Ben Azariah (Missoula, 1977)
Zeb.	=	Zebaḥim
Zuckermandel	=	M. S. Zuckermandel, ed., Tosephta, based on the Erfurt and Vienna Codices (Trier, 1881-1882)

TRANSLITERATIONS

א	=	'
ב	=	b
ג	=	g
ד	=	d
ה	=	h
ו	=	w
ז	=	z
ח	=	ḥ
ט	=	ṭ
י	=	y
כ ך	=	k
ל	=	l
מ ם	=	m
נ ן	=	n
ס	=	s
ע	=	c
פ ף	=	p
צ ץ	=	ṣ
ק	=	q
ר	=	r
שׁ	=	s
שׂ	=	ś
ת	=	t

INTRODUCTION

I. The Subject of the Tractate

Mishnah Tractate Maaser Sheni deals with second tithe. This tithe is one of the agricultural gifts which Israelite farmers are obligated to set aside from their annual harvest.[2] Mishnah's authorities identify it with the agricultural offering mentioned in Deut. 14:22-26 (cited below). There Scripture describes an agricultural levy which Israelite farmers are to set aside each year and eat in Jerusalem.[3] Mishnah Maaser Sheni, for its part, spells out how this tithe is to be handled until it is properly eaten in the holy city. As I shall explain, the tractate's particular concern is to protect the tithe from loss or misuse.

Let us begin our discussion by considering the relevant passage from Scripture.

> You shall tithe all the yield of your seed, which comes forth from the field year by year. And before the Lord your God, in the place which he will choose, to make his name to dwell there, you shall eat the tithe of your grain, of your wine, and of your oil, and the firstling of your herd and flock; that you may learn to fear the Lord your God always. And if the way is too long for you, so that you are not able to bring the tithe, when the Lord your God blesses you, because the place is too far from you, which the Lord your God chooses, to set his name there, then you shall turn it into money, and bind up the money in your hand and go to the place which the Lord your God chooses, and spend the money for whatever you desire, oxen or sheep, or wine or strong drink, whatever your appetite craves; and you shall eat there before the Lord your God and rejoice, you and your household.

According to Deuteronomy, each year Israelite farmers must set aside one-tenth of their harvest. This tithe is to be taken to Israel's central sanctuary and eaten there.[4] The author of Deuteronomy takes account of the burden this requirement places on farmers who live far from the central shrine. To facilitate their pilgrimage, the Deuteronomist allows these farmers to sell the produce which they have designated as second tithe and in its place to bring money to the city. This money is then used in the city to purchase food which is eaten in place of the original produce. Scripture goes on to say that the tithe, either the original produce or what has been brought to replace it, is "eaten before the Lord." While the meaning of this phrase is unclear, it does indicate that the second tithe is to be eaten in the central sanctuary.[5]

For Mishnah, as for Scripture, the common folk of Israel are responsible for designating produce to serve as second tithe, bringing this tithe to Jerusalem and eating it there. What is of special interest to Mishnah's authorities is that this tithe is set aside to be eaten only in Israel's holy city. This fact means, in their view, that it is holy. It is for this reason that Mishnah is interested in second tithe.[6] Since the produce is holy, it must not be treated in the way the farmer treats unconsecrated produce, which is available for his everyday use in any place. Rather, M.'s authorities want to establish special restrictions to govern precisely how produce in this consecrated status may be used. The tractate before us spells out these restrictions.

Mishnah's authorities want to insure that the farmer who designates produce as second tithe does not lose or destroy any part of it through carelessness. The full amount of what he sets aside must be eaten in Jerusalem. The tractate develops this concern in two ways. The first is to define the specific uses the farmer may make of the produce. He must eat the produce and not use it in some other way, such as an unguent. The second, and more important concern, is that the farmer receive the full value of his consecrated produce when he sells it. If he receives an insufficient value in coin, he will end up eating in Jerusalem less second tithe than he originally separated. If he sells two dinars' worth of consecrated produce for only one and one-half dinars in coin, for example, only three quarters of the value of the produce is transferred to the money. The remaining value of the second tithe is still in the produce, which now belongs to the buyer. This creates two problems. On the one hand, the farmer will not buy in Jerusalem the proper value of produce to be eaten as second tithe. On the other hand, the buyer has produce which, unknown to him, is still in a consecrated status. He will use this produce as though it were unconsecrated, thus violating the sanctity of second tithe. It is important, therefore, that the selling price of the consecrated produce accurately reflect its value, so that the full value of second tithe is passed to the coins.

Now that the principal concerns of the tractate are clear, let us explain how it is constructed. As we shall see, it unfolds in three units. The first describes restrictions which apply to the use of produce and coins in the status of second tithe. The second unit takes up the topic of greatest concern to M.'s authorities, permissible uses of consecrated produce and coins. This unit is concerned in particular with exploring how the farmer transfers the status of second tithe from one item to another, such that none of its value is lost. The last unit takes up topics of secondary concern. It first considers produce which is in a status analogous to that of second tithe, namely fruit growing on a tree or vine during its fourth year of growth. This fruit is comparable to second tithe in that it is both holy and eaten by the farmer himself in Jerusalem (Lev. 19:23-25). The tractate concludes with a block of material which deals with the law of removal (Deut. 14:27-29). This law declares that every three years the farmer must properly distribute all tithes which he has separated but not given to its appropriate recipients. If he cannot properly use the produce, he must destroy it. This is important as regards second tithe since after the Temple's destruction, second tithe could not be eaten.

With the structure of the tractate in mind, let us turn to an outline of each of its parts. This will allow us to see how the individual pericopae of the tractate have been organized into the essays described above.

I. Improper Disposition of Second Tithe (1:1 - 1:7)

A. Improper use of consecrated food (1:1 - 2F)

1:1 Produce in the status of second tithe may not be sold, given as a pledge or used as a counterweight, but it may be given as a gift.

1:2A-F These same rules apply to tithe of cattle.

B. Improper use of consecrated coins (1:2G - 7)

1:2G The farmer may not consecrate coins which are defaced, out of circulation or inaccessible.

1:3-4 The farmer may not purchase with consecrated money inedible items such as hides or jugs unless these are an intrinsic part of the food they accompany.

1:5A-C The farmer may not purchase edible items that are not foods, such as water or salt.

1:5D-6 The farmer may not purchase foods outside Jerusalem.

1:7 The farmer may not purchase items which he cannot eat, drink nor use as a lotion.

The redactor opens the tractate with a series of restrictions which apply to produce or coins in the status of second tithe. In this way, he indicates that second tithe is unlike unconsecrated produce and thus introduces us to the topic of the tractate as a whole. The unit first deals with produce which has been declared as second tithe (A) and then considers money which is brought to Jerusalem in place of that produce (B). This second unit begins with a definition of what constitutes proper money (M. 1:2G) and goes on to describe how this money is to be used (M. 1:3-6). Two interpolations have been added to this material. M. 1:2A-F, which deals with tithe of cattle, has nothing to do with second tithe. It is placed after M. 1:1 apparently because it applies the same rules to another agricultural levy. The final pericope, M. 1:7, repeats the point already made at M. 1:3-4. It is included in the tractate in order to form a transition to the material which follows in Unit II.

II. Proper Disposition of Second Tithe (2:1 - 4:12)

A. Proper use of consecrated food (2:1-4)

2:1A-H Consecrated food must be used in its normal way.

2:1I-P If consecrated food and unconsecrated food are mixed together, the value of the consecrated food is determined by its proportion in the mixture.

2:2 Simeon and sages dispute over the proper use of oil.

2:3-4 An appendix to the foregoing in which the Houses dispute the proper use of fenugreek and vetches.

B. Proper transference of the status of second tithe to and from coins (2:5 - 2:9)

1. From coins to coins (2:5 - 2:9)

2:5 Consecrated and unconsecrated coins mixed together: coins randomly picked from the batch are deemed consecrated in place of the original money.

2:6 Consecrated and unconsecrated coins mixed together: the status of the consecrated coins is transferred to a fresh batch of coins.

2:7 Houses dispute: may the farmer exchange silver coins for gold coins?

2:8 Transfer of status outside Jerusalem.

2:9 Transfer of status inside Jerusalem.

2. <u>From coins to produce</u> (2:10 - 3:4)

2:10 Consecrated coins may be spent a little at a time.

3:1-2 The farmer may not purchase as second tithe produce in the status of heave-offering, since this is in a consecrated status.

3:3 The status of second tithe may be transferred from one person's coins to another person's produce.

3:4 The status of second tithe may be transferred from a farmer's coins to his own produce.

3. <u>From produce to coins</u> (3:5 - 4:8)

a. <u>After the produce has been in Jerusalem (3:5 - 3:13)</u>

3:5 Produce brought into Jerusalem no longer may be sold.

3:6 Houses' dispute: does this rule apply to produce which is unprocessed?

3:7-8 Appendix on determining the boundary of Jerusalem.

3:9 Houses' dispute: does this rule apply to consecrated produce rendered unclean?

3:10-11 The application of this rule to food purchased for use as second tithe and which becomes unclean.

3:12-13 The purchase in Jerusalem of inedible items.

b. <u>Before the produce has been in Jerusalem</u> (4:1-8)

4:1 Produce is sold at local market prices.

4:2A-D Produce is sold at the lowest prevailing price.

4:2E-G Produce, the price of which is unknown, is auctioned.

4:3 An added fifth of the food's value is paid when the farmer transfers the status of second tithe to his own coins.

4:4-5	An appendix on how the added fifth legally may be avoided.
4:6	What if the agreed-upon selling price is different from the produce's market value?
4:7	Does selling consecrated produce require an oral declaration?
4:8	Coins are spent at the current value they have at the time of sale.

c. Produce and coins the status of which is in doubt (4:8-12)

4:9	Found coins are deemed unconsecrated unless there is specific evidence to the contrary.
4:10-11	Found produce is deemed unconsecrated unless its container is marked in a specific way.
4:12	Found coins are deemed consecrated only if they are found in a place where consecrated coins are known to have been.

This unit is a fine piece of redactional work, having a clear beginning, middle and end. It opens, appropriately, with a discussion of consecrated produce (A). This introduces the next logical topic, and the one of major concern, namely coins which are consecrated in place of that produce (B). The concluding section (C) deals with doubts, a typical redactional technique for ending a major thematic unit. The redactor has arranged his materials, then, in such a way that his discussion unfolds in a logical manner. Let us now consider each of these subunits in turn.

A makes a simple point. Consecrated food must be used in its normal way. Since this section is short, and deals with its subject in a cursory manner, it is clearly of only minor concern to the redactor. He has placed it here in order to introduce his real interest, which is taken up in B.

B considers the transfer of the status of second tithe from one object to another. Its point is that, as we said earlier, second tithe must suffer no loss in value on account of the transaction. The unit is elegantly arranged, consisting of three essays, each dealing with one of the transactions possible with consecrated goods. These are the exchange of consecrated coins for other coins (B/1), second, the use of these coins in Jerusalem to buy produce (B/2), and third, the sale of consecrated produce for coins (B/3). This is not the order in which the farmer would effect the transactions, since the sale of produce (B/3) is logically prior to the other transactions, in which coins are deconsecrated. By arranging the material in this order, however, the redactor ends the unit with the most substantive of the three essays. This essay itself is divided into two subunits, each dealing with one of the two possible locations of the produce when it is sold: inside or outside Jerusalem. By so dividing the material, the redactor deals first with situations in which second tithe may not be sold at all (3a) and, then, with the laws governing the sale in cases when it is allowed (3b).

C closes the unit, as we said, by considering cases of doubt. The essay deals first with coins, then with produce and concludes by returning our attention to coins. It is organized, then, in the pattern of a-b-a, a fine redactional flourish ending the tractate's major essay.

III. Special Topics (5:1-15)

A. Produce of a planting's fourth year (5:1-5)

5:1 Fruit of the fourth year must either be eaten in Jerusalem or must be sold, as is true of second tithe.

5:2 Originally fruit of the fourth year was sold only if it grew further than a three day's journey from Jerusalem. Nowadays, it is sold irregardless of where it grows.

5:3 Houses' dispute: Do other laws which apply to second tithe apply as well to fruit of the fourth year?

5:4-5 How the price of such fruit is established in special cases.

B. The law of removal (5:6-15)

5:6 All agricultural gifts must be distributed by Passover of the fourth and seventh years or they must be destroyed.

5:7 Houses' dispute: How is the removal carried out nowadays?

5:8 Are crops still in the field liable to the law of removal?

5:9 Produce may be distributed through an oral declaration.

5:10-14 The text of the confession (Deut. 26:13-15) is the subject of a midrashic interpretation, which finds in it reference to the rabbinic laws of agricultural gifts.

5:15 A catalog of five legal actions of Yoḥanan, the High Priest. The first of these is his abolition of the recitation of the confession.

With his examination of second tithe completed, the redactor turns to two related topics. The first, A, concerns produce growing on a tree or vines in its fourth year after planting. Such produce is like second tithe in that it is both holy and is eaten by the farmer in Jerusalem. M. declares that the same considerations which apply to the handling of second tithe apply as well to such produce. The second block of material, B, deals with the final disposition of food the farmer has consecrated but which he is unable to use. The produce, in M.'s view, must be destroyed. This discussion is especially suitable as the conclusion to our tractate. This is because after Jerusalem's destruction, second tithe could not be eaten. This concluding material, then, explains how second tithe is to be handled in the period after 70 C.E.

II. Method of Exegesis

My analysis of Mishnah Maaser Sheni begins with a fresh translation of the tractate. The purpose of the translation is to enable the reader to understand the exegetical problems posed by each pericope's content, formulation and syntax. My translation differs from earlier ones, then, primarily in that it attempts faithfully to replicate the syntax and patterned language of the original Hebrew. To make formal analysis possible, I divide each pericope into its smallest literary units, labelling them with the letters of the alphabet. In this way, I am able to discover the structure of each pericope. At the same time, I endeavor to render the Hebrew into intelligible English so that its meaning is clear. In order to do this, and yet preserve the literary character of the original, I have found it necessary to add words or phrases not supplied in the Hebrew but clearly assumed by the author. I have placed such added language in square brackets. The reader is thus made aware of the original wording of the Hebrew while at the same time having before him a readable text. I have systematically checked my Hebrew text against manuscript variants, calling the reader's attention to these variant readings which are important for the interpretation of the pericope. I have also checked each pericope against parallels in the talmudic literature. For the reader's convenience, I include with each translation a list of citations noting the occurrence of the same phrases or sentences elsewhere in Mishnah, the Tosefta or in the Talmuds.

The exegesis is meant to describe the point of law which lies at the heart of each pericope or small unit of tradition. I want to know, in each case, what meaning the formulator himself intended his words to convey. To accomplish this purpose, I begin by considering the unit's formal traits. This is so because it is by means of their formulation of the law that the authorities of M. draw our attention to the issues which primarily concern them. This kind of analysis is possible because of the highly patterned language used in M. By choosing to express their point in one pattern rather than another, the formulators indicate where their interest in each pericope lies. They may, for example, present two balanced rules which contain matching, but opposed, apodoses. By so framing matters, the formulator draws our attention to the contrast between the apodoses and thus to the issues which produce the contrast. In my exegesis, I identify this critical tension and explain the logical basis for the two opposing rules. A second common literary technique used by M.'s formulators is the dispute. This consists of a superscription followed by opposing apodoses or a rule followed by an opposing gloss. The point of the pericope emerges in each case through the contrast between the two opinions. From the positions articulated in the dispute, we can infer the premises that both parties hold in common as well as the gray area which the common premises produce and which gives occasion for the dispute. As before, my exegesis identifies the common premises and describes the logic which stands behind each opinion. A third formal trait is the list, a catalog of different items or actions grouped together because, in the formulator's mind, they share some common trait. The problem is to locate this unstated characteristic so as to identify how the formulator interprets the diverse facts before him. My job as exegete is to discover that common trait. In each case, then, my interpretation of the pericope is guided by its formal traits. These lead me to the central concern of each unit and thus to the point of law the formulator himself wants to articulate.

Once I have identified the main point of concern in each pericope, I turn to other rules in the pericope, asking whether these serve the pericope's main point or are only tangentially related to it, either introducing the pericope or illustrating its material. I thus am able to give an account of the full structure of each pericope and to describe the logic by which it has been put together. In this way, I claim to make clear the full meaning of each unit of material.

Having described the meaning of each pericope, I step back and ask about the point of the tractate as a whole. To do this, I begin by considering how each pericope is thematically related to the material around it. This analysis allows me to identify thematic units within the tractate. The results of this analysis are presented in the introductions to each of the tractate's chapters as delineated in the printed editions. With the basic thematic units identified, I describe how these have been organized so as to form the tractate as a whole. From this I discover the central issues or concerns which motivate the redactor to put together our particular tractate. The results of this effort are found in Part I of this introduction.

Although my study has its own particular purpose and method, as I have explained, I regularly draw upon the work of classical exegetes to the tractate. I consult these earlier commentaries both for help in understanding difficult words and phrases and for a sense of the logical possibilities inherent in Mishnah. Their work is relevant to mine because these earlier writers are interested, as am I, in discovering the logic behind M.'s law and in articulating that logic. At the same time, the use I make of classical exegesis is limited because of the different purposes of our commentaries. My purpose, as I said, is to understand what M.'s laws meant to the people who created them. The task classical commentators set for themselves, to the contrary, is to show how M.'s rules fit in and complement rabbinic law in general. This is so because they conceive of rabbinic law as an all-encompassing and consistent whole. Their work, then, is one of harmonization and synthesis, in which M.'s laws are understood and interpreted in the context of the full corpus of rabbinic law. Thus, while I can draw on classical exegetes for help in understanding points of language and logic, I often disagree with them as regards the context and meaning of that law. In my commentary

I have taken care to draw the reader's attention to points at which I depart from classical interpretations of a pericope, and to give my reasons for rejecting the classical view in favor of my own interpretation.

It remains for me to discuss my translation and exegesis of Tosefta. I have included this document in my study because of its close relationship, substantive and formal, to M. As my exegesis of Tosefta shows, this document is closely linked to M., relying on M. both for its agendum of issues and for its redactional order. It frequently cites M. directly or supplements what M. has to say with further material. It also complements M.'s discussions by introducing new materials. Tosefta thus explores possibilities left open by M. and broadens M.'s own investigations into points of law. This it does through use of the same formalized language and patterned syntax of M. For these reasons, I provide a translation and exegesis of it along with my commentary to M.

For my translation and exegesis of T. I use essentially the same procedure as I do for M. (Tosefta Maaser Sheni appears here in English for the first time.) The principal difference is that my primary aim as regards T. is to make clear its relationship to M. To this end, I underline in my translation all passages in which T. directly cites M. I also provide the reference to M. in square brackets following the translation. This allows the reader quickly to see the points of contact between M. and T. The purpose of my commentary is to understand in what way T. qualifies or expands Mishnah's law.

My translation of M. depends primarily on the text published by Ḥanoch Albeck and Ḥanoch Yalon.[7] This text has been checked against the variant readings listed in the critical edition of Mishnah Zera^cim edited by N. Sacks.[8] Important variant readings have been indicated in my translation by the use of parentheses. I have also made regular use of English translations of M., primarily those of H. Danby[9] and M. H. Segal.[10] For T., I rely on the text published by S. Lieberman.[11] He includes in his edition variant readings and, occasionally, his own suggested emendations to the text. Where these are important for understanding T., I include these in my translation in parentheses. I also cite other relevant emendations, such as those of Samson of Sens and GRA.[12]

CHAPTER ONE
MAASER SHENI CHAPTER ONE

Produce designated as second tithe either must be taken to Jerusalem for consumption or must be sold and its value in coin used to purchase other food as second tithe in the holy city. The chapter before us spells out the implications of these facts. In particular, it develops the notion that the farmer must use the produce only as food, and not for any other purpose. The point of the law is that since the produce and coins are consecrated, the farmer is not free to use them in any way he chooses. Rather, he must use them only as Scripture specifies, that is, as food. The chapter's discussion unfolds in two thematic units, dealing first with prohibited uses of produce (1:1-2) and then with prohibited uses of coins (1:3-7).

M. 1:1 lists four types of transactions for which the farmer may not use produce in the status of second tithe. In each of the cases listed, the farmer receives some monetary benefit from using his consecrated produce for something other than food. By prohibiting such transactions, M. makes it clear that the farmer enjoys only limited control over how he may use consecrated goods. M. 1:2 carries this principle forward, applying it to tithe of cattle. Like second tithe, this animal is not fully the property of its owner. It follows that it may not be used in business transactions which will financially benefit the farmer.

M. 1:2G completes this unit by considering the one business transaction in which consecrated produce may be used. This is when the farmer sells produce in the status of second tithe in order to use the proceeds to purchase other food in Jerusalem. M. makes the obvious point that if the produce is sold in this way, the farmer must receive in exchange only coins which can be used in Jerusalem. This is to insure that the farmer will indeed eventually eat second tithe in Jerusalem.

Having discussed the use of consecrated produce, the chapter investigates the use of consecrated coins. The redactor here stresses two points: that the coins must be used to purchase food (M. 1:3-5C, 7) and that the coins must be used in Jerusalem (M. 1:5D-6). We consider first what is to be done if the farmer uses his consecrated coins to purchase non-foods.

M. 1:3-4 declare, first, that in some cases the purchase of inedible items is in fact legitimate. This is so if the farmer incidentally acquires certain inedible items along with food, items such as animal skins or jugs. In these cases, the inedible items do not constitute a purchase in their own right since the farmer has no intention of purchasing these particular items. Consequently, they are not deemed to take on the status of second tithe from the coins. M. 1:4-5C and M. 1:7 next investigate cases in which the farmer clearly does purchase inedible items with consecrated money. The discussion now concerns how the farmer rectifies matters. Two theories are offered. According to M. 1:4 and 1:7, the purchased items themselves have become consecrated. That is, the transaction effects the deconsecration of coins even though the sale is illegal. The purchases must be resold, therefore, and the realized money used to purchase food in their stead. M. 1:5A-C holds, to the contrary, that no transfer of status takes place since the sale is in all events prohibited. The purchases remain unconsecrated while the paid-over coins are still holy. The farmer, then, must retrieve his consecrated money and return the purchases.

The second misappropriation of consecrated money investigated by M. concerns spending the money outside of Jerusalem (M. 1:5D-6). Of interest here is the farmer's intention when he made the purchase. If he spent the coins inadvertantly, being unaware that he was paying with consecrated coins, the transaction is deemed to have had no effect. That is so because the farmer has not intended to deconsecrate the coins. The purchases remain unconsecrated, therefore, and the farmer is obligated to bring these back to the merchant. He then retrieves in return the original coins, which remain consecrated (M. 1:5A-C). If he knowingly paid over consecrated money, however, the transaction does effect a transfer of consecration. The farmer now intentionally deconsecrated the money. It follows that his purchases take on the status of second tithe and must be taken to Jerusalem for consumption.

1:1

A. [As to produce in the status of] second tithe —
B. (1) They do not ('yn) sell it,[1] (2) and they do not take it as a pledge (mmšknyn), and (3) they do not give it in exchange [for other produce to be eaten as second tithe].
C. And they do not (wl') reckon weight with it.
D. And in Jerusalem, a man may not say to his friend: "Here is wine for you, now give (wtn) me oil."
E. And [this rule applies] likewise to all other [consecrated] produce.
F. But ('bl) they give [it] to each other as a gift.

M. 1:1 (C: y. M.S. 1:1
F: M. M.S. 3:1)

Produce declared second tithe is taken to Jerusalem and is eaten there by the farmer (Deut. 14:22-27). M. points out that the farmer is in fact prohibited from making any other use of the produce (A-C). That is, once produce has been consecrated, it may be used only for its declared purpose, namely as food to be eaten in Jerusalem by the farmer. The reason is that the farmer does not enjoy exclusive rights to the use of the food. Rather, he must use it in accordance with the restrictions laid down by Scripture. It follows that the farmer may not use such produce for his own personal benefit, as a means of earning money, for example (B), or for his convenience (C).

D-F define more precisely the type of exchange that B(3) prohibits. The point emerges from the contrast between D and F. In D, the farmers give each other produce with the expectation of receiving some favor in return. In this case, the transaction is like a sale in that each farmer receives some benefit. In F, to the contrary, each farmer gives the other a gift without expecting to receive something in return. This type of transaction is allowed, even if each party ends up receiving produce from the other.

A. [As to] second tithe:
B. (1) they do not sell it, (2) and they do not take it as a pledge, (3) and they do not give it as a pledge, (4) and they do not exchange it.
C. And they do not reckon the weight of golden dinars with it [M. 1:1A-C]

D. even to deconsecrate [other] second tithe with them [i.e., with the golden dinars].

E. And he may not give them [that is, coins in the status of second tithe] to a money-changer to derive benefit from them or (w-) to loan them out (wlhlwwtn; E reads: l' ylwh 'wtm) to gain stature through them (lht^{c}ṭr; GRA emends to: lht^{c}šr).[2]

F. [But] if [he does so] in order that they not rust (y^{c}lw ḥlwdh), it is permitted.[3]

G. How [do we interpret] "they do not sell it" [=B(2)]?

H. One should not say to another, "Here are 200 [zuz worth of consecrated produce], now give (wtn) me one maneh [i.e., 100 zuz in unconsecrated coin]."

I. How [do we interpret] "they do not take it as a pledge" [=B(2)]?

J. He should not enter his [the borrower's] house and take his second tithe as a pledge.

T. 1:1 (p. 243 , ll. 1-6)

K. How [do we interpret] "they do not give it as a pledge" (=T. 1:1B(3))?

L. One should not say to another, "Here is this tithe, let it be yours (wyh' bydk; E reads: wh' bydk), so give [me] unconsecrated [money or produce] for it."

M. How [do we interpret] "they do not exchange it" [=B(4)]?

N. One should not say to another, "Here is wine for you, now give me oil" [M. 1:1D].

O. But he says to another, "Here is wine for you because I have no oil," and the other one says, "Here is oil for you because I have no wine."

P. It turns out that they are (nmṣ'w) exchanging and [yet] not exchanging, [rather] they are doing a favor for one another.

T. 1:2 (p. 243 , ll. 6-10)
(B: b. Shab. 22b, y. M.S. 1:1; E, F: T. B.M. 4:2; J, O: y. M.S. 1:1)

T. is divisible into three units: A-D, which cites and expands M. 1:1A-C; E-F, which discusses coins which were used to deconsecrate produce in the status of second tithe; and G-P, a commentary on T. 1:1B. D glosses C. F limits the rule of E. The third section systematically cites T. 1:1B, at G, I, K and M, and provides a specific illustration of each. The four question-and-answer units of this third unit are cast in the same pattern: kṣd xx (= a citation from T. 1:1B) plus l' plus a singular imperfect participle. D glosses N and is explained by P.

In C-D produce (or coins) in the status of second tithe are used to weigh golden dinars, which themselves will be used to deconsecrate other produce in the status of second tithe. In this case, the farmer receives no benefit from this use of consecrated produce since he is simply preparing other consecrated foods. Nonetheless, his use of consecrated produce is prohibited.

E directs our attention to coins which have been used to deconsecrate produce in the status of second tithe. Such coins take on the status of second tithe from the produce. Consequently, they become subject to the restrictions which once applied to the produce and may not be used by the farmer to gain benefit. It follows they may not be given to a money-changer, who will derive benefit from them by appearing to be rich (so MB), nor may they be loaned out, because the lender will realize some gain through the transaction. The coins are to be used only for their declared purpose, namely to be used in Jerusalem to purchase food to be eaten as second tithe.

H gives us an example of a prohibited sale. The farmer here sells his consecrated food cheaply, indicating he expects the purchaser to take the food to Jerusalem and eat it there. As we noted in M., however, the farmer must himself eat his consecrated produce in Jerusalem and may not transfer this obligation to others. Consequently, this type of sale is prohibited.

N-P spell out the point M. makes in M. 1:1D-F. N's illustration of prohibited exchanges repeats the language of M. 1:1D. O then describes the kind of exchange that the law will allow, the mutual giving of gifts (M. 1:1F). P explains why O's transaction is acceptable.[4]

A. [As to] produce in the status of second tithe:[5]

B. they do not put it on a callous [Jastrow: on the sole of the foot] and not on a lichen[6] (ḥyt; D: zwyt; A: zzyt; Lieberman amends to ḥzzyt).

C. And they do not make it [into] an amulet.

T. 1:3 (p. 243, ll. 10-11)

The formulary pattern of this pericope is familiar from M. 1:1 and T. 1:1. The question now is whether or not consecrated produce or coins may be used for some benefit other than monetary gain.[7] As we see, even such benefit is prohibited. These rules are perfectly in accord with M. and T.'s notion that produce or coins in the status of second tithe must be used only for consumption in Jerusalem.[8]

1:2

A. [As to] the tithe of cattle:

B. (1) [the farmers] do not sell it [when the animal is] unblemished [and] alive; (2) and not [when the animal is] blemished, [whether it is] alive or slaughtered.

C. (3) And they do not give it as a token of betrothal to women.

D. [As to] the firstling [i.e., the first calves of the year's herd]:

E. (1) they [i.e., the priests] sell it [when the animal is] unblemished [and] alive; (2) and [when the animal is] blemished, [whether it is] alive or slaughtered.

F. (3) And they give it as a token of betrothal to women.

G. They do not deconsecrate [produce in the status of] second tithe with (1) a poorly minted coin ('symwn)[9] nor with (2) a coin that is not [currently] circulating (hmṭb^{c} š'ynw ywṣ'), nor with (3) money that is not in one's possession (hm^{c}wt š'ynn bršwtw).

M. 1:2 (D-F: b.B.Q 12b,
b.Tem. 7b, y.Qid. 2:5;
G: b.B.M. 47b)

The pericope restates the principle, introduced at M. 1:1, that the farmer does not have the right to use consecrated food as though it were his own personal property. The point is made by contrasting the laws applying to tithe of cattle (A-C) with those applying to firstlings (D-F). The farmer is not free to use consecrated tithe of cattle for his personal benefit while the priest may use firstlings in any way he wants. Since the latter offering is not consecrated, no restrictions apply to its use until it is offered at the altar.

Tithe of cattle is that one-tenth of the herd which the farmer must consecrate each year (Lev. 27:32-33). Like produce in the status of second tithe, such animals must be eaten by the farmer in Jerusalem. The farmer then does not have the right to use these beasts as he would use unconsecrated animals in his possession. Such animals therefore may not be used in any business transaction from which the farmer expects to gain personal benefit (B-C). In contrast, firstlings are not consecrated but are deemed to be the personal property of the priests (M. Bekh. 4:1). Since these are the property of the priests, it follows that the priestly owner may use these animals for his own personal benefit (D-F).

The pericope is complicated by the fact that B and E introduce the categories of blemished vs. unblemished and alive vs. slaughtered. This is especially a problem at B since we expect the law to prohibit the use of tithe of cattle for the farmer's personal benefit irregardless of the animal's physical state. As we shall see, however, these criteria do make sense in connection with firstlings. We can account for the presence of these criteria in B, then, on the grounds that the redactor wants to produce a perfectly balanced doublet. Let us now consider briefly how these criteria relate to firstlings.

E(1) indicates, first, that the priest may sell living firstlings but not slaughtered ones. The reason is that a priest who slaughters a firstling must sprinkle its blood on the altar and offer its fatty parts as a burnt-offering (Nu. 18:15f). That is, once the priest slaughters an animal, he is no longer free to dispose of it as he will. It follows that while a living animal may be sold, a slaughtered one may not.[10] E(2) now declares that this distinction between alive and slaughtered is irrelevant if the firstling is blemished. Blemished animals are in all events unfit for the altar. Since the priest cannot offer up the blood and fatty parts of the blemished animal, it follows that he is free to sell it even if it is slaughtered, as E(2) declares. We see, then, that the criteria mentioned in E are in fact appropriate for firstlings. As we stated above, they were introduced at B in order to create a balanced pericope.

G introduces a fresh topic, coins which a farmer may receive when selling produce he has designated as second tithe. These coins take on the sanctity of the consecrated produce and are used in Jerusalem to buy other food to be eaten as second tithe. It follows that the farmer must be sure that the coins he uses for this purpose are proper coins. That is, he must be sure to receive coins which will be acceptable to grocers in Jerusalem. G defines for us the characteristics of such proper coins. The coins must be properly minted (G1), generally accepted (G2), and immediately available to their owner (G3).

A. They do not deconsecrate [produce in the status of] second tithe with a poorly minted coin [M. 1:2G].

B. And R. Dosa permits.

C. But [regarding] small coins (prwṭwt kṭnyt; E.D: prwṭwt kṭnywt), which are given [as] a token in the bathhouse, all agree that they do not deconsecrate it [i.e., the second tithe] with them.

T. 1:4 (p. 243, l. 11 - 244, l. 2) (b.B.M. 47b; y.M.S. 1:2 A-B: M. Ed. 3:2)

A cites M. 1:2G(1). R. Dosa's conflicting opinion in B is narrowed by C. The small coins referred to in C are tokens minted by private individuals rather than the government.[11] Since they are not money in the strict sense of the term, they may not be used to deconsecrate second tithe produce.

A. They do not deconsecrate it [i.e., produce in the status of second tithe],

B. (1) not with a coin from the Revolt [of Bar Kokhba],
(2) and not with a coin that is not [currently] circulating,
(3) and not with money that is not in one's possession [M. 1:2G].

T. 1:5 (p. 244, ll. 13-14) (y.M.S. 1:2)

C. How [do we interpret "coin from the Revolt" and "coin that is not circulating]?"

D. [If] one had coins [of the Revolt of Bar] Koziba[12] (mᶜwt kzbywt) and coins of Jerusalem[13] (mᶜwt yrwšlmywt) they do not deconsecrate it [i.e., produce in the status of second tithe] with them.

E. If he did deconsecrate [with them], they [i.e., the coins] have not acquired [the status of second] (qnw: E: qnh = he has not acquired) tithe.

F. But with a coin which circulates on the authority of (mšwm; E: lšwm = in the name of) the earlier rulers (mlkym hr'šwnym), they deconsecrate it [i.e., the second tithe] with it.

G. They do not deconsecrate it,

H. either with local [i.e., Palestinian] money (m^{c}wt škn = money from here) in Babylonia, or with money of Babylonia (Lieberman supplies kn = here).

I. But if he was here and he had [Babylonian] coins in Babylonia, they deconsecrate it [i.e., second tithe] with them.

J. How [do we interpret], "They do not deconsecrate with money that is not in his possession [= M. 1:2G(3)]?

K. [If] one had coins hidden in a [Roman] camp (qṣṭr')[14] or in the royal hill country (wbhr hmlk),[15]

L. or his wallet (kysw) fell into a cistern,

M. even if he knows it is there,

N. they do not deconsecrate it [i.e., produce in the status of second tithe] with them.

O. And if he did deconsecrate [second tithe], he has not acquired [the status of second] tithe [for the money].

T. 1:6 (p. 244, ll. 14-20)
(C-H: b.B.Q. 97ab; H-K: y. M.S. 1:2; K: b.B.Q. 98a)

T. gives specific instances of types of coins which are not considered to be proper money. Coins from the revolt of Bar Kokhba and Jerusalemite coins (D) were minted by illegitimate authorities. On that account, they are not deemed to be proper money. F gives the contrary case, coins which, although out of date, were minted under proper authority. These are valid coins. G-I specify that coins not in circulation refer to coins of one country which are not accepted in another (H). Coins of a foreign country may be used to deconsecrate produce in that foreign country, however, even if the owner is elsewhere (I). This leads us to the third consideration (J-O): cases in which the coins are not accessible to the owner at all. The laws assume the situation obtaining in Palestine after the Bar Kokhba revolt when Roman troops were stationed throughout Palestine and when Jerusalem had been converted to Aelia Capitolina. The rationale behind deconsecrating second tithe produce with money is that the coins are taken to Jerusalem and used to procure other produce. If the owner is unable to get the coins to Jerusalem because they are inaccessible, they obviously cannot be used to deconsecrate second tithe.

A. Similarly (kywṣ' bw) [Lieberman: just as he may not redeem second tithe produce with coins that are in a Roman camp or in Jerusalem or in a cistern...],

B. one who is walking down the road with copper coins in his hand and sees a thug ('ns) coming toward him, should not say, "Lo, those are deconsecrated with the copper coins that I have in my house."

C. But (w; E and D omit) if he did so, his words have effect (kyymyn).

T. 1:7 (p. 244, ll. 21-22)
(b.B.Q. 115b)

D. [If] one was carrying jugs of wine and jugs of oil and saw that they were broken ('wtn šnšbrw), he should not say, "Lo, these are made heave-offering and tithes for the produce I have in my house."

E. If one did so, lo, it [the produce in his house] is forbidden [to be eaten because it is still in the status of untithed produce].

T. 1:8 (p. 244, l. 22 — 245, l. 24)

In A-C the thug is about to take the sanctified money that the man has in his hands. The man quickly tries to deconsecrate the coins with other coins he has in his house. In this way, he hopes to prevent the sanctified coins from falling into the hands of the thug, who will certainly use them for secular purposes. Normally, one may not transfer the status of second tithe from one set of coins to other coins stamped on similar metal (M. 2:6-9). T. allows the transaction to be considered valid in this case in order to insure the proper use of the second tithe coins.

D-E present a contrasting case. Now the man wants to sanctify the produce he is about to lose. If he is able to sanctify the leaking oil or wine, he will have fulfilled his legal obligation to separate tithes without having to forfeit more produce. The leaking produce is lost to him anyway. Here the intention is not to safeguard sanctity, but to fulfill the law as cheaply as possible. Since the man makes the declaration knowing that he cannot properly carry the tithing procedure through, T. does not accord validity to his statement.

GRA and Lieberman both want to emend T. 1:7 in light of T. 1:8.[16] According to them, the question is one of deconsecrating what is at home with the coins the man is carrying with him and which he is about to lose. The assumption, as in D-E, is that the man will try to sidestep the law in some way. That is, the exegetes have the man transferring the sanctity of coins at home to the coins he will soon lose, thus being relieved of the responsibility of taking them to Jerusalem. According to these emendations, the coins in his hand do indeed become sanctified, presumably because the coins are under his control when he makes his declaration.

1:3-4

A. (1) One who purchases [in Jerusalem] a domesticated animal (bhmh) [with money in the status of second tithe] for [use as] a peace-offering,
or, ('w, most MSS: w-: and) a wild animal hyh for [use as] ordinary meat (lbśr t'wh),[17]
B. the hide is deemed to be unconsecrated (yṣ' ... lḥwlyn)
C. even though [the value of] the hide exceeds [the value of] the meat.
D. (2) Sealed jugs of wine [which are purchased in Jerusalem with consecrated money] —
E. in a place where they are normally sold (šdrkm lmkwr)[18] sealed,
F. the jar is deemed to be unconsecrated.
G. Nuts and almonds [which are purchased in Jerusalem with consecrated money] — their shells are deemed to be unconsecrated.
H. Wine [made from grape skins and stalks][19] (tmd) —
until it has fermented, it is not bought (nlqḥ) with money [in the status of second] tithe.
I. After it has fermented, it is bought with money [in the status of second] tithe.

M. 1:3 (H-I: y. Ma. 5:5)

1:4

J. (1) One who purchases [in Jerusalem] a wild animal [with money in the status of second tithe] for [use as] a peace-offering,
(most MSS: w-: and) a domesticated animal for [use as] ordinary meat,
K. the hide is not deemed to be unconsecrated [i.e., it is in the status of second tithe and must be resold]. (Several MSS add: even though [the value of] the hide exceeds [the value of] the meat.)
L. (2) Open or sealed jugs of wine [which are purchased in Jerusalem with consecrated money] —
M. [in] a place where they are normally sold open,
N. the jar is not deemed to be unconsecrated.
O. Baskets of olives (several MSS: figs) and baskets of grapes with their container [purchased with consecrated money] — the value of the container is not deemed to be unconsecrated.

M. 1:4 (K-M: b. Men. 82a)

A farmer purchases in Jerusalem with consecrated money certain inedible items, such as skins of animals or jugs containing wine. At issue is whether or not these items, which cannot be eaten, become consecrated as second tithe on account of the sale. If they do become consecrated, they must be resold so that food can be purchased in their place, since the farmer is required to eat second tithe. M.'s point, as we shall see, is that consecration adheres to only those inedible items which the farmer purposely acquires with his consecrated money. This is so because he has intentionally transferred the status of second tithe to them. If, however, the farmer acquires the hides or jugs only incidentally along with food that he purchases, the inedible items are not deemed to be in a consecrated status. The farmer has not intentionally bought these, and so the status of second tithe is not deemed to have been transferred to them. M. makes its point in a series of three contrasting pairs: A-C and J-K, concerning animal skins, D-E and L-N, regarding wine jugs, and finally the singletons at G and O. We will consider each pair in turn.

In A-C a domesticated animal is purchased for use as a peace-offering, or a wild animal, not suited for the altar, is purchased as ordinary meat. These animals become consecrated with the status of second tithe. They are therefore subject to regulations regarding second tithe produce, but their hides remain unconsecrated (B) and may be used as the farmer sees fit. This is so because the farmer does not intend to purchase the hides themselves, but rather acquires them only incidentally along with the animals. He may dispose of these hides as he would any other unconsecrated item. J-K reverse the situation. Now the farmer purchases with consecrated money a domesticated animal for use as ordinary meat, or a wild animal for use as a peace-offering. Both of these sales are improper, however, since domesticated animals purchased with consecrated coins may not be used as ordinary meat nor may wild animals be used as an offering.[20] Since the animals have been purchased with money in the status of second tithe, however, they do become consecrated. Consequently, the animals must be resold so as to consecrate money used to purchase proper produce. In these cases, M. declares the hides to be sanctified along with the rest of the animal. This insures that the entire animal is resold and its full value used to purchase other food.

From animal hides, M. moves to a consideration of the parallel case of wine bought with consecrated money. The question concerns the status of the jug which contains the wine. If the wine is normally sold in sealed jugs (D-F), the container is understood to be an intrinsic part of the sale. It is thus analogous to the hides of properly purchased animals, and is not deemed to acquire sanctity. The farmer may dispose of the jug as he will. If the wine is normally sold in open containers (L-N), it is possible to buy the wine without its jar. Now the jar does constitute a purchase in its own right and so is deemed to acquire the status of second tithe. It must be resold so that other produce may be purchased in its stead.

G and O apply the same logic to produce. Nutshells are an intrinsic part of the fruit and therefore do not constitute a distinct purchase. It follows that they do not acquire the status of second tithe and may be discarded. Fruit baskets (O), however, are not naturally a part of the food they contain. If purchased with consecrated coins, they must be resold.

H-I introduce a new issue. Its point is that only items in the status of food may be purchased with consecrated money, a theme taken up in the following pericopae (M. 1:5-7). Temed wine is a cheap beverage made by soaking grape skins and stalks in water and letting the mixture ferment. Before fermentation begins, the mixture is not deemed to be a food item and so is not to be purchased with consecrated money. The fermented liquid is deemed to be a kind of wine and so may be purchased for use as second tithe.

A. They buy a wild animal and a fowl for use as ordinary meat, but not for use as peace-offerings [= M. 1:3-4].
B. When they decreed that ordinary meat would make hands unclean, they ruled, "They do not buy a wild animal for use as ordinary meat,"
C. But they buy fowl for use as ordinary meat.
D. R. Leazar ben Judah from Abilene (<u>'yš 'blyn</u>)[21] says, "Not even fowl [is bought] for use as ordinary meat.
E. For the grape-pressing (<u>gt</u>) made by [one who ate] ordinary meat is unclean in connection with Holy Things (<u>lqwdš</u>), unclean (D, E: clean) in connection with heave-offering."

T. 1:9 (p. 245, ll. 24-28)
(cf. T. Niddah 9:18)

T. is best understood in light of the more detailed tradition of T. Niddah 9:18:

A. At first ordinary meat was regarded as insusceptible to uncleanness.
B. They reverted to decree concerning it that it should be subject to uncleanness of the hands.
C. They reverted and decreed concerning it that it should be subject to uncleanness [also] through contact.
D. They reverted and decreed concerning it that it should be like carrion itself and impart uncleanness [also] through carrying.
E. They reverted and said every olive-pressing made in connection with ordinary meat is unclean for Holy Things but clean for heave-offering.
F. They reverted and ruled that this applies to a beast and not to fowl.
G. R. Eleazar ben Judah of Kefar 'Ublin said, "Even if it was made in connection with one fowl or in connection with one chicken, it is unclean for Holy Things and clean for heave-offering."[22]

Two issues have been conflated into our account: the uncleanness of ordinary meat (A-B+E), and the question as to whether or not fowl is considered to be meat for this purpose (C+D). A restates the situation assumed by M. 1:3A. Then comes the decree of B whose consequences are illustrated at E: one who eats ordinary meat becomes unclean and can render liquids (i.e., the grape juice) unclean. Purchasing a wild animal for ordinary meat with consecrated money is prohibited, therefore, because all second tithe food must be eaten in a state of cleanness.[23]

A. One who acquires a cask [of wine] in Jerusalem [with money in the status of second tithe] —
[if the cask is] open, he must eat [produce] in place of [the value of] the vessel;
[if the cask is] sealed, he does not need to eat [produce] in place of [the value of] the vessel.
B. Under what circumstances?
C. For [casks] of (bšl) wine.
D. But for those of (1) brine, and of (2) vinegar, and of fish-brine (mwrys),[24] and of (4) oil, and of (5) honey,
E. whether [the cask] is open or sealed,
F. he does not need to eat [produce] in place of [the value of] the vessel.
G. [As for] the broom-shaped twig of the date-palm (dytb'wt šl tmrh),[25]
H. and (1) wicker-baskets (pwṭly'wt) and (2) pomace baskets of dates,
I. one does not eat [produce] in its place.
J. If they were [ordinary] baskets,
lo, this one eats [produce in place of the value of the basket].
K. And [as to] all other things,
one does not eat [produce] in their place,
L. (1) either in place of kernels (hḥrṣnyn), (2) or in place of husks (hzgyn), or in place of mash (hgpt).[26]

T. 1:10 (p. 245, 11. 28-34)
(B-D: T. M.S. 2:18)

T. is divided into three units: A-F, dealing with casks of liquids; G-J, types of baskets; and K-L, refuse of produce. A unfolds in accordance with the principles of M. 1:3D-F. The sealed cask is intrinsic to the purchase of the wine it contains. Since it is not deemed to be a distinctive purchase, it does not become sanctified by the consecrated money used to purchase it. An open cask, however, is considered to constitute a separate purchase, since the wine may be sold directly out of the container. Now if the cask is purchased with consecrated money, it does take on the sanctity of second tithe. It must be resold so that consumables may be purchased in its stead. This general statement is then limited by B-F. T. distinguishes between casks of wine and casks of other liquids. Jugs which contain the items listed in D cannot be used for any other purpose since the produce in them either cannot be fully removed or leaves an unpleasant odor. As a result, the jugs must be sold along with their contents and so are not deemed to take on the status of second tithe from the coins.

The general theme of the foregoing discussion is carried on in G-J. Now our attention is directed to various types of baskets used for dates. Produce is hard to remove from these baskets because of the basket's construction (G, H(1)) or because the produce has been mashed into it (H(2)). In such cases, the basket is deemed inseparable from its contents and so does not constitute a separate purchase.

K-L restate the principle of M. 1:3G, the status of nutshells. The inedible refuse normally acquired with produce does not become sanctified when the produce is purchased with consecrated coins.

A. One who acquires a deer in Jerusalem,
does not need to eat [produce] in place of the hide.
B. [If the deer is] slaughtered,
he must eat [produce] in place of the hide.

T. 1:11 (p. 245, ll. 34-35)

A wild animal is purchased in Jerusalem for use as ordinary meat. M. 1:3A-C declares that in this case the hide is not consecrated. This is the point of A. B introduces a new consideration. The farmer purchases an animal which is already slaughtered. Now the meat surely can be purchased apart from the hide. In such a case, the hide does constitute a separate purchase and so is consecrated.

A. They do not rent houses in Jerusalem [to those bringing offerings] because they [i.e., the houses] are [the property] of [all the] tribes.

B. R. Lazar b. R. Simeon says, "Also not ('p l') beds."

T. 1:12 (p. 245, l. 35 — p. 246, l. 37)
(B. Yoma 12a, Meg. 26a)

C. The hides of sanctified animals —
innkeepers come and take them by force.

T. 1:13a (p. 246, l. 37)
(b. Yoma 12a, Meg. 26a)

A-C concern payment for room in Jerusalem for people coming to the city to eat second tithe. Since places in Jerusalem cannot be rented, it was customary for the visitors to leave something behind for the innkeeper. According to T., hides from peace-offerings, declared to be unconsecrated in M. 1:3A-C, are normally left behind for this purpose. So strong was this custom that innkeepers could come and take the hides by force if the boarder was not willing to leave it on his own.

M. M.S. 1:5-6

A. One who buys [with money in the status of second tithe] (1) water, or (w-) (2) salt, or (3) pieces of fruit attached to the ground,

B. or ('w) pieces of fruit which cannot reach Jerusalem,

C. has not acquired [the status of second] tithe [for these items].

D. One who buys [outside Jerusalem with money in the status of second tithe] pieces of fruit:

(1) unintentionally (šwgg) [i.e., not realizing the coins were consecrated] —
let their payment (dmyhn) be returned to its [former] place [i.e., to the purchaser who bought them by mistake];

(2) on purpose (mzyd) —
let [the pieces of fruit] be brought up and eaten in the [holy] place [i.e., Jerusalem].

E. And if the Temple does not exist ('yn mqdš),
let [the pieces of fruit] rot.

M. 1:5

F. One who buys [outside Jerusalem] a domesticated animal [with money in the status of second tithe]:

(1) intentionally —
let its payment return to its [former] place;

(2) on purpose —
let [the animal] be brought up and eaten in the [holy] place (z: bkl mkwm).

G. And if the Temple does not exist,
let it be buried with its hide.

M. 1:6
(b. Qid. 45b-46a)

M. consists of two formal units: A-C and a doublet at D-G. The phrase, "One who buys" (hlwqḥ), appearing at A, D and F, continues the pattern of M. 1:3 and 1:4. Again the issue is whether or not the status of second tithe is transferred to items improperly purchased with consecrated money. We deal with two cases, first with the purchase of items the farmer cannot eat in Jerusalem (A-C) and second with the purchase of edible food outside Jerusalem (D-G).

A-C deal with items which are not proper foods (A) or foods which will not be edible by the time that the farmer gets them to Jerusalem (B).[27] In either case, what the farmer has brought cannot be eaten as second tithe in Jerusalem.[28] The result, M. claims, is that the status of second tithe has not been transferred from the coin. This is so because the transaction itself is prohibited and thus has no legal effect. The purchased items are still unconsecrated while the coins remain sanctified. In order to rectify matters, the farmer must recover the coins and spend them properly.

In D-G the farmer purchases edible produce (D-E) or animals (F-G), but does so improperly by buying them outside of Jerusalem. Whether or not the transaction has effect depends on the farmer's intention. If the purchase was made unintentionally, that is, if the purchaser did not realize he was using consecrated coins, the transaction has no effect, as was true in A-C. He must return the produce or animal, which remains unconsecrated, and receive back his sanctified coins. If he knowingly spent consecrated coins, however, the status of second tithe has been transferred. The farmer now has intentionally deconsecrated sanctified coins. These purchases may not be sold again but must themselves be taken to Jerusalem and consumed there. It is the farmer's intention when paying over the coins, then, which effects the transfer of status. After the destruction of Jerusalem, when food in the status of second tithe cannot be consumed, the farmer has no choice but to let such food rot.

A. (1) They do not redeem [i.e., sell, produce in the status of] second tithe in Jerusalem in [add with Lieberman hzh; the present] age [i.e., after the destruction of the Temple],
B. (2) and they do not separate [produce as] second tithe in Jerusalem in the present age,
C. (3) and they do not deconsecrate [coins in the status of] second tithe in Jerusalem in the present age,
D. (4) and they do not sell [produce in exchange for coins in the status of] second tithe in Jerusalem in the present age,
E. (5) and they do not take [produce in the status of] second tithe out of (reading with E: m) Jerusalem in the present age [so that it can be sold outside the city; cf. M. 3:5];
F. and if he took [it] out [of the city], lo, [the produce] [is left to] rot [i.e., it may not be sold].

T. 3:13 (p. 260, ll. 44-48)
(cf. T. San. 3:6)

A. [If] they deconsecrate in Jerusalem coins [in the status of second tithe] with produce in the present age —
B. the House of Shammai say, "[Both] this and that [i.e., the coins and the produce] [are in the status of] second tithe."
C. The House of Hillel say, "The coins [retain their status] and the produce [retains] its status."

T. 3:14 (p. 260, ll. 48-50)
(y. M.S. 1:4)

T. 3:13A-E lists five rules. F is explicitly linked to E(5) but, as we shall see, refers to A-D as well. T. 3:14 goes over the ground of T. 3:13C.

After the destruction of the Temple, produce in the status of second tithe no longer could be eaten (M. 1:5-6). In light of this fact, T. declares illegitimate all transactions in the destroyed city involving such produce. The farmer may neither separate second tithe in Jerusalem from untithed produce (B) nor may he buy produce in Jerusalem to be used as second tithe (C-D). Furthermore, produce in the status of second tithe which is brought into the city may not be sold (D) nor may it be taken out of the city again (E). The farmer who brings produce in the status of second tithe to Jerusalem after the destruction has no choice but to let it rot, as M. 1:5E and 1:6G have stated.

T. 3:14 has the Houses dispute the consequences of a farmer's attempting to buy in the destroyed city produce for use as second tithe. The Shammaites declare that the coins the farmer spends are not deconsecrated, since the sale itself is prohibited. Nonetheless, the produce he purchases does enter the status of second tithe. In this way the farmer is penalized for his misdeed.[29] The Hillelites disagree. Since the transaction is not allowed to begin with, they hold that no change in status occurs on its account. The coins remain in their consecrated status and the produce remains unconsecrated. This same view is expressed in M. 1:5A-C.

D. Sprouts of the service-tree (lwlby hzrdyn)[30] and the carob tree — before they have sweetened (E, D add: are not) purchased with money [in the status of second] tithe. After they have sweetened, they are purchased with money [in the status of second] tithe.

E. But Colocasia (lwp)[31] and mustard and lupine,

F. and all other preserved vegetables,

G. whether after they have sweetened or before they have sweetened, are purchased with money [in the status of second] tithe.

H. Safflower seeds (bnwt hry^{c}; D: ḥry^{c}; E: hdy^{c})[32] are purchased with money [in the status of (Lieberman + D: second)] tithe, but they do not receive uncleanness as food.

I Safflower cakes —[33]

J. Said R. Yosé, "R. Yoḥanan ben Nuri came to ('ṣl) R. Halafta [and] said to him, 'As to safflower cakes, what is the rule concerning their being purchased (mh hn šylqḥw) with [money in the status of second] tithe?' He said to him, 'They are not purchased [with such money].' He said to him, 'I also rule thus, but Aqiba has ruled they are purchased [with consecrated money].'"

T. 1:13b (p. 246, ll. 38-43)
(D-G: T. Uq. 3:9; H: M. Uq. 3:5; I: y. Ḥag. 3:4

K. Palm tree pith (qwr)[34] is purchased with [money in the status of second] tithe but does not receive uncleanness as food.

L. Unripe date berries (kpnywt)[35] are purchased with [money in the status of second] tithe and do receive uncleanness as food.

M. R. Judah says, "Palm tree pith — lo, it is like wood in all regards except that it is purchased with [money in the status of second] tithe."[36]

N. "Unripe date berries — lo, [they are] like pieces of fruit in all regards except that they are exempt from tithes."[37]

O. R. Simeon said in the name of R. Leazar, "Saffron (kwrkwm)[38] is not purchased with [money in the status of second] tithe because there is no flavor in it; rather [it is] for decoration."

P. Said R. Judah ben Gadish before R. Leazar, "My father's employees used to sell brine in Jerusalem [for consecrated money]." They said, "Who said [this], for maybe they sold fish with it [M. 1:5]?"

T. 1:14 (p. 246, ll. 44-247, l. 49) (J-M: T. Uq. 3:10-11, b. Er. 28b; L-M: M. Uq. 3:7; L: y. Sheb. 4:6; O: b. Er. 27a)

I discern three units in T.'s discussion: D-G, H-O and P. The point of the first unit is made by the contrast between D and E-G. The young sprouts at D are not deemed to be foods until they sweeten and so are not purchased for use as second tithe until that time. The items at E-G are deemed to be foods even if they have a bitter taste and so are properly purchased with consecrated money. T.'s second unit (H-O) discusses various types of produce the status of food of which is in doubt. First we have items which are not considered to be foods even though they are edible (H-K). Being edible, they may be purchased with consecrated money, although they are not held to be susceptible to uncleanness as food. Next we consider items, such as unripe berries in L, which are deemed to be foodstuff. These, of course, are purchased with consecrated coins and also are susceptible to uncleanness as foods. Finally, O, we have garnishes which are not considered to be foods at all. Consequently, they are neither purchased with consecrated money nor do they become unclean as foods. In the last unit (P), T. reads the law of M. 1:3-4 into M.'s prohibition against purchasing salt or water with consecrated coins (M. 1:5A). Brine, being a mixture of salt and water surely falls in this category. If the householder intends to purchase only the fish which is soaking in the brine, he may spend consecrated coins for the entire jug, even though he acquires brine along with the fish.

A. One who purchases a domesticated animal for a peace-offering [with second tithe money outside of Jerusalem] by mistake [not realizing that the coins were consecrated], let its payment return to its [former] place; [if he did so] on purpose, let it be brought up and eaten in the [holy] place [= M. 1:6F].

B. [If this takes place] at the present time, let [the animal be left to] die [= M. 1:6G].

C. Said R. Judah, "Under what circumstances? When he intended to purchase it for [use as a] peace-offering to begin with. But if he intended to deconsecrate [second] tithe money [improperly by purchasing the animal for use as ordinary meat], whether by mistake or on purpose, let the payment return to its [former] place."

D. If it was a blemished animal, whether [it was purchased with consecrated money] by mistake or on purpose, let the payment return to its [former] place.

E. If they were unclean pieces of fruit, whether [they were purchased with consecrated money] by mistake or on purpose, let the payment return to its [former] place.

T. 1:15 (p. 247, ll. 50-55)
(C: b. Qid. 56a)

A-C read the issue of M. 1:3-4 into the law of M. 1:5-6. The farmer has purchased the animal for a legitimate use, but he has done so outside of Jerusalem. The law of M. 1:5-6 applies. If the farmer intentionally misspent his sanctified money, the purchase is effected and the now-consecrated animal must be consumed in Jerusalem. If, however, the farmer was unaware that he was using consecrated coins, the animal remains unaffected by the sale. The farmer must return the animal and his still-consecrated money is refunded. Judah's elaboration in C applies to the opposite case. The farmer purchases the meat for an improper purpose to begin with (cf. M. 1:4J-K). Since this purchase will not allow the farmer properly to fulfill his second tithe obligation, Judah rules that no transfer of sanctity takes place at all. The animal is returned and the money refunded, whether or not the farmer was aware that he had used sanctified coins. This, as we have seen, is the Hillelite view in T. 3:14 and applies as well if the farmer purchases a non-food, such as water or salt (M. 1:5A-C).

D-E make a related point. Blemished animals and unclean produce are not properly eaten as second tithe. If the farmer purchases such foods with consecrated coins, he in no way can fulfill his obligation. Consequently, the sale is deemed to have no effect as regards the transfer of the status of second tithe to the food, as was the case in C.

1:7

A. They do not purchase (1) male slaves, or (2) female slaves, or (3) real estate
B. (4) or an unclean animal,
C. with second tithe funds (dmy).
D. And if he purchased [one of the above], let him consume in its stead [its same value in other produce].
E. They do not purchase (5) the sacrificial birds of zabim, nor (6) the sacrificial birds of zabot, nor (7) the sacrificial birds of women who are unclean after childbirth,[39]
F. (some MSS lack:) (8) or sin-offerings, (9) or guilt-offerings
G. with second tithe funds.
H. And if he purchased [one of the above], let him consume in its stead [its same value in other produce].
I. This is the general rule: Anything that is not suitable for eating, drinking or anointing which is [purchased] with money [in the status of] second tithe, let him consume in its stead [its same value in other produce].

M. 1:7
(b. Qid. 56a; A-H:
M. Sheb. 8:8)

The pericope consists of two balanced units, A-D and E-H, with their general rule articulated in I. The material here continues M. 1:5A-C's list of items which are not to be purchased for use as second tithe. In each case, the listed item cannot be used as food by the farmer. The first three are not foods at all. Unclean animals, while edible (4) are unfit for use as consecrated food. The remaining exampla (5-9) are eaten by priests and not by the person who brings them. If one of these prohibited items is purchased with sanctified coins, the farmer is obligated to purchase an equivalent value in edible produce and to consume the new produce in Jerusalem.

CHAPTER TWO
MAASER SHENI CHAPTER TWO

There are two ways in which the farmer can fulfill his obligation to eat in Jerusalem produce declared second tithe. He may, first of all, carry the designated produce itself to Jerusalem and consume it there. Alternatively, he may transfer the status of second tithe from the designated produce to coins and take these to Jerusalem in place of the produce. Since coins are less bulky than produce, such a procedure greatly facilitates his pilgrimage. In Jerusalem the coins must be used to purchase a second batch of produce. This batch then enters the status of second tithe and is eaten as such, just as if it was the original batch. In the chapter before us, M. deals with the question of the proper disposition of these two items, produce and coins, which have the status of second tithe. M. 2:1-4 first discusses restrictions on the use of produce designated as second tithe. M. 2:5-9 is concerned with the transfer of the status of second tithe from coins to other coins. This type of transaction is of interest because Scripture makes no provision for it. M. allows the exchange, however, on condition that it be conducted in a way analogous to the sale of produce. One rule is that the transfer must move from an item of lesser intrinsic value to one of higher intrinsic value. That is, just as the farmer gives fruit for metal, so may he give coins of cheaper metal, such as copper, for coins of better metal, such as silver. The last pericope of the chapter deals with the exchange, in Jerusalem, of coins in the status of second tithe for produce.

M. 2:1A-C adduces the basic principle which governs the use of produce in the status of second tithe. Such produce must be consumed in Jerusalem in the way produce of its type is normally used. That is, the farmer must eat what is normally eaten, drink what is normally drunk and anoint with what is normally used as a lotion. This prevents the farmer from wasting or in some other way misusing, what is consecrated. M. 2:1F-H supplies an obvious adjunct to this principle, ruling that nothing may be added to produce in the status of second tithe which will prevent the produce from being used in its normal way. Spices may not be added to oil, for example, if they render the oil inedible.

In line with this notion, M. 2:1/I-P assume that unconsecrated ingredients may be mixed with produce designated second tithe if this does not change the normal use of the consecrated produce. This material goes on to address a second issue: how we determine the value of the consecrated portion of the mixture. This is a problem if the farmer wants to sell the dish and finds that it is worth more than the sum of the cost of its ingredients. K-P rule that in such a case we reckon the monetary value of the consecrated ingredient on the basis of that ingredient's proportion to the whole.

M. 2:2-4 are an appendix to this unit. They address themselves to the principle stated at M. 2:1A-C, that produce declared second tithe is used as food, drink or lotion, as appropriate. Simeon and sages now dispute whether or not anointing is indeed a proper mode of using such produce. Their debate in M. 2:2 is followed by supplementary material at M. 2:3-4. This added material serves two functions. First, it supplies laws which are needed to understand the debate at M. 2:2. At the same time, by establishing the normal use of vetches and fenugreek, it casts the debate in M. 2:2 into the terms of M. 2:1's concern that consecrated produce be used only in its usual way.

The second major unit of our chapter, M. 2:5-9, addresses two issues which naturally arise once we assume that the status of second tithe may be transferred from produce to coins and, thence, from the original coins to some other coins. The first is introduced at M. 2:5-6E. Consecrated and unconsecrated coins have become confused in a single batch. Under discussion is whether the status of second tithe continues to adhere to the specific consecrated coins that were lost in the batch (M. 2:5), or whether we consider the status of second tithe to be dispersed uniformly throughout the mixture (M. 2:6). The second concern appears in M. 2:6F - 2:9. These pericopae spell out the kinds of coins to which the status of second tithe may be transferred. The rule is that outside of Jerusalem, the status of second tithe is always transferred from coins stamped on inferior metal to coins stamped on better metal, from copper to silver, for example. Inside Jerusalem, the procedure is reversed, allowing the farmer to receive small change when buying food in the city. In this way the exchange of coins for other coins is made to parallel the exchange of produce for coins. In each case the status of second tithe is transferred to items of greater material worth.

The chapter's final pericope, M. 2:10, takes up the issue of how consecrated coins are exchanged in Jerusalem for produce to be eaten as second tithe. In the case before us, the farmer purchases produce but intends to eat only part of it as second tithe. At issue is whether or not the farmer can split up a single purchase into consecrated and non-consecrated parts. M. rules that this is possible. The farmer may stipulate that only that part of the purchase which is actually consumed as second tithe is to take on that status.

2:1

A. Second tithe is permitted for eating and drinking and anointing;

B. for eating that which is normally used for eating,

C. for anointing with that which is normally used for anointing.

D. One may not anoint with wine or vinegar.

E. But he anoints with oil.

F. They do not spice oil [in the status of second tithe which has been purchased as a food]

G. nor do they buy spiced oil with money [in the status of] second tithe [for use as a food].

H. But he spices wine [in the status of second tithe].

I [If unconsecrated] honey or spices fell into [consecrated wine] and increased [its value] (hšbyḥw)

J. the increase is [divided] proportionately [between the wine and the honey or spices].

K. Fish which were cooked with leeks [in the status of] second tithe (qplwṭwt),[1] and they [the leeks] increased [the fish's value] —

L. the increase is [divided] proportionately.

M. Dough [in the status of] second tithe which he baked and increased [its value] —

N. the increase is [accounted entirely] to the [dough designated as] second [tithe].

O. This is the general rule: Any [ingredient] whose benefit (šbḥw) [to the mixture] is perceptible (nkr) — the increase [in value] (hšbḥ) is [divided] proportionately.

P. And any [ingredient] whose benefit [to the mixture] is not perceptible — the increase [in value] is [accounted] to the second [tithe].

M. 2:1 (A-B: T. Ter. 9:10; A-D: M. Sheb. 8:2; F: M. Ter. 11:1; J-K: y. Ter. 10:1)

The pericope consists of three units (A-C, D-H and I-P) all of which develop a single point. Produce in the status of second tithe must be used only in the way produce of its type customarily is used. That is, once the farmer consecrates produce, he is obligated to use it in the way deemed proper for that type of commodity. This basic rule is stated explicitly at A-C. The following two units explore the application of this rule when the consecrated food is mixed with other ingredients, such that it is no longer in its pure, natural, state. As we shall see, such mixtures are deemed permissible if the farmer still uses the consecrated food in its customary way.

D+E first discuss the special case of oil. Oil is unusual in that it is customarily used both as a food and as an unguent (D-E). Consecrated oil thus can be used in one of two ways. If the farmer specifically prepares the oil for one purpose, such as for an unguent, however, he must then use it only for that purpose. Spicing oil, for example, gives it a fragrance which makes it into a desirable unguent. Oil purchased as a food may not be spiced, since this would make it into a lotion (F). Conversely, such spiced oil, which is prepared as a lotion, may not be purchased as a food (G). Spices may be added to wine, however, since this adds flavor to the wine but does not tempt the farmer to anoint with it (H).[2]

We turn in I-P to cases in which unconsecrated ingredients are added to consecrated foods in order to prepare a fancy dish. According to what we have just said, such mixtures are valid as long as the consecrated food is still used in its customary way. Such mixtures create a problem, however, if the farmer wants to sell the prepared dish and take the proceeds to Jerusalem to be used to purchase second tithe. The problem is to determine what proportion of the money the farmer receives for the dish must be taken to Jerusalem. We consider three examples. In the first case (I-J) produce designated as second tithe has been improved by the addition of unconsecrated spices. When the food is sold, the amount of money to be taken to Jerusalem depends on the proportionate cost of the consecrated main ingredient to the cost of the unconsecrated relishes. Maimonides offers the following example: the consecrated produce is worth one selaᶜ while the unconsecrated ingredients are worth three selaᶜs. The mixture, however, is worth five selaᶜs, a one-selaᶜ increase over the cost of its ingredients. We now account 1¼ selaᶜs' value to the second tithe in the dish and 3 3/4 selaᶜs to the unconsecrated portion. As a result, 1¼ selaᶜs of the purchase price of the dish becomes imbued with the sanctity of second tithe.

The second example reverses the proportion of the ingredients. Now we have an unconsecrated main dish embellished with consecrated spices. The same principle is applied and the value of the various ingredients is determined by proportion.

Our third example addresses the value of work, as opposed to physical ingredients, added to consecrated material. In the case before us, consecrated dough is baked into loaves with a resulting increase in value. M. rules that none of the increase in value is ascribed to the overhead expenses of baking. The reasoning is articulated in O-P. Since the work is not physically present in the finished product, the entire selling price the farmer receives for the bread is deemed consecrated.[3]

A. They do not soak dates in order to produce date beer from them,

B. and [they do] not [soak] raisins to produce raisin juice,

C. but they crush them and make them into puree (ṭrym').[4]

D. Concerning spices, it is permitted [to soak them] because this is their [normal] mode of preparation (ml'ktn).

E. [If one] ties the spices [in a bundle] and sets them in a cooking dish,

F. if their flavor is dissipated, they are permitted [to be used in this way]

G. and if [their flavor is] not [dissipated] they are prohibited [from being used in this way].

T. 2:2 (p. 249, ll. 11-14)
(D-F: T. Ter. 9:7; T. Sheb. 6:6-7)

T. offers two exemplifications of M.'s rule that one must use consecrated produce in its customary way. The first contrasts the use of fruit with the use of spices. In A-B, the farmer proposes to soak consecrated dates and raisins and to ferment the resulting liquid in order to produce a kind of beer. Dates and raisins, however, are not normally prepared in this manner. Consequently, processing them in this way is prohibited. In D, the farmer wants to soak spices in order to add flavor to a dish he is cooking. This is allowed since this is the normal way in which spices are used. C adds a second contrast to the law of A-B. Now the farmer wants to mash the consecrated food into a puree. In contrast to the liquid of A-B, this puree will have in it the entire bulk of the fruit. This kind of preparation may be eaten, presumably, because the consecrated fruit, although mashed up, is still fully eaten as a food.[5]

E-G are concerned with the proper use of spices purchased with consecrated coins. We want to know whether the spices may be tied up in a kind of bundle. The point of the contrast is that if the bundling up interferes with the spices' ability to flavor a cooking dish, it is prohibited. This is so because the spices now can no longer serve their normal function.

A. A *log* of wine [in the status of second tithe] into which fell a *log* of [unconsecrated] honey

B. and lo, [the mixture] contains two *logs* —

C. the increase [in value] is [divided] proportionately (cf. 2:1H-I).

D. A cooked dish of [consecrated] produce which was seasoned with unconsecrated spices —

E. the increase [in value] is [divided] proportionately.

F. But (*w-*) [a cooked dish of] unconsecrated produce which was seasoned with spices [in the status of] second tithe —

G. its second tithe [ingredients] may not be released [from their sanctified status] through redemption (*m^{c}śr šny šlw bpdywn*; Lieberman: *m^{c}śr šny šl' bpdywn* — may not be released without redemption).

T. 1:16 (p. 247, ll. 55-57)
(y. M.S. 2:1)

T. is composed of three units: A-C, D-E and F-G. The first two units illustrate the rulings of M. 2:1/I-P. Any increase in value due to a mixing of unconsecrated and consecrated produce is divided proportionately between the two types of ingredients. The third unit addresses itself to a somewhat different issue. We now have a dish made up of unconsecrated produce to which some consecrated spices have been added. The spices have totally dissolved into the dish and thus have lost their discrete identity. We want to know whether or not these spices, which no longer exist, can be deconsecrated. According to the manuscript reading, these spices may not be redeemed. This is so, MB suggests, because the spices lose their independent existence when they dissolve in the broth. Since the consecrated portion of the mixture may not be redeemed, the entire dish must be taken to Jerusalem and consumed there.

Lieberman wants to amend the text to allow the redemption of the consecrated ingredients. He holds that since the spices were added into the dish, they are present in it, even if we can not find them. Like all other added ingredients, these spices can be deconsecrated.

A. Second tithe and unconsecrated leeks (qplyt) which were mixed with each other —

B. the increase [in value] is [divided] proportionately.

T. 1:7 (p. 247, ll. 58-59)

Now we have two batches of the identical spice mixed together. Should there be an increase in value, it will be divided between the consecrated and unconsecrated spices in accordance with the rulings of M. 2:1.

A. They do not make wine into lotion ('lntyt)[6] or oil into a [spiced] mixture (ᶜrb).

B. If he made wine into a lotion or oil into a [spiced] mixture,

C. he anoints with the oil but he does not anoint with the [lotion made from] the wine or vinegar.

D. for oil is normally [used] for anointing [but] wine and vinegar are not normally [used] for anointing.

T. 2:3 (p. 249, ll. 14-16)
(T. Ter. 9:7)

A restates the general principle of M. 2:1A-H. Consecrated produce must be used only in its normal way. If wine is turned into a lotion, it follows, the lotion may not be used for anointing. This is not the normal use of wine. Similarly, oil, which is normally eaten, should not be mixed with spices to produce a fragrant ointment. If the oil is mixed in this way, however, the resulting lotion may still be used for anointing. This is so because such use is still consistent with the normal use of oil.

A. [If] one had dough of [second] tithe [wheat] and baked it —
B. the increase [in value] is [divided] proportionately [cf. M. 2:1M].
C. [If] the bread fell apart,
D. [or] the dish was burnt (<u>nqdḥ</u>) —
E. the decrease [in value] is [accounted] to the second [tithe].
F. [If] one sorts out wheat and grinds it and sifts it,
G. it is sufficient if he redeems it at the market price of [unrefined] wheat.
H. R. Yosé says, "In a place where they figure in the cost [of baking],
I. "they redeem the wheat but they do not redeem the loaves [of bread] [at the market price of wheat].
J. "In a place where they do not figure in the cost [of baking],
K. "they redeem even the loaves [at the market price of wheat]."
L. Oil [in the status of] second tithe and clumps of unconsecrated spices which were mixed together —
M. the increase [in value] is [divided] proportionately.
N. "This is the general rule," said R. Yosé,
O. "Everything which adds to the volume [and] the benefit [of which] is perceptible (<u>šbḥw nkr</u>) — the increase is [divided] proportionately.
P. "[If] it does not add to the volume, [even though] the increase [to the value of the mixture] is perceptible — the increase [in value] is [accounted only] to the second [tithe]."

T. 1:18 (p. 248, l. 58 — 249, l. 65)

The pericope consists of four units: A-E, F-K, L-M and N-P. A-B restate the case introduced at M. 2:1M, but arrives at a different conclusion. The increase in value produced by baking the dough is divided proportionately between the value of the consecrated dough and the unconsecrated work. That is, overhead expenses incurred in preparing the dough is treated by T. as is all other ingredients added to consecrated foods. It receives its fair share of an increase in value which occurs on its account. This is not the case, however, if the prepared dough suffers a loss in value (C-E). In this case, we do not apportion the loss equitably between the work and the dough. Overhead expenses must in all events be paid. The result is that if there is a loss in value, it is borne entirely by the consecrated portion.

F-K investigate further steps in the production of bread. We begin with the processing of wheat. Normally, when wheat is refined, no increase in its value occurs. Since the worked wheat is sold at the same price as unworked wheat, we do not take into account when selling consecrated wheat the value of unconsecrated labor which went into its processing. The entire amount of money received for the loaf is consecrated. When the wheat is baked into bread, we take local business practices into account. If baked goods are normally sold at a higher price than a comparable amount of flour, we regard the extra value as reflecting the work put into the preparation of the bread. The extra value, consequently, is divided proportionately between the overhead and the dough. If the bread is sold at the market price of wheat, however, there is no increase in value. In this case, we do not worry about according part of the price of consecrated loaves to overhead expenses, and the entire proceeds are in the status of second tithe.

L-M is predictable from what M. has given us. The increase in value enjoyed when two ingredients are mixed together is apportioned between the two ingredients.

O-P restate the general principle of M. 2:1/O-P. As we recall, M. 2:1 stipulates that any increase in value accruing to a dish on account of its being prepared or cooked is divided proportionately between its consecrated and unconsecrated ingredients. This is true any time that the presence of both types of ingredients is discernible. T. now qualifies M.'s rule, limiting it to cases in which there has also been an increase in overall volume. The point, then, is that only if both types of ingredients are physically present in the mixture do we divide any increase in value between them.

2:2-4

A R. Simeon says, "They do not anoint in Jerusalem with oil designated as second tithe."

B. But the sages permit [them to anoint with oil declared second tithe].

C. They [sages] said to R. Simeon, "If [the ruling] is lenient in regards to heave-offering, which is subject to a more stringent rule [by allowing priests to anoint with it], should we not rule leniently for second tithe, which is subject to a less stringent rule [and allow the farmer to anoint with it]?

D. He [R. Simeon] said to them, "No, (mh l'; other MSS read mh or l'), if [the ruling] is lenient in regards to heave-offering, which is subject to a more stringent rule, in a situation (mqwm) where it was lenient in regards to vetches (kršynym)[7] and fenugreek (tltn),[8] should we be lenient in regards to second tithe in a situation where [the law] was not lenient in regards to vetches and fenugreek?

M. 2:2

A. Fenugreek which is [in the status of] second tithe must be eaten [when] sprouting (ṣmḥwnym)[9] [and is preserved in cleanness].

B. And [fenugreek which is declared] heave-offering —

C. the House of Shammai say, "Anything done with it is [done] in [a state] of cleanness, except for shampooing with it."

D. And the House of Hillel say, "Anything done with it is [done] in [a state of] uncleanness, except for soaking it."

M. 2:3
(cf. T. Uq. 3:13-14)

A. Vetches which are [in the status of] second tithe must be eaten [when] sprouting.

B. But they enter Jerusalem and come out [i.e., once they have been brought into Jerusalem they may be taken out again].

C. [If] they become unclean —

D. R. Tarfon says, "Let them be divided up among [pieces of] dough,"

E. and sages say, "Let them be deconsecrated (ypdw)."

F. And [as for vetches which are declared] heave-offering —

G. the House of Shammai say, "They soak and crush (<u>spyn</u>)[10] [them] in [a state of] cleanness, and they feed [them to cattle] in [a state of] uncleanness,"

H. and the House of Hillel say, "They soak [them] in [a state of] cleanness and crush [them] and feed [them to cattle] in [a state of] uncleanness."

I Shammai says, "Let them be eaten dry."

J. Rabbi ᶜAqiba says, "Anything done with them is [done] in [a state of] uncleanness."

M. 2:4
(F-J: M. Ed. 1:8)
(cf. T. Uq. 3:13-14)

Simeon and sages dispute what M. 2:1E has taken for granted — that anointing is a proper mode of consuming consecrated oil. Their dispute at M. 2:2 A-B is followed by a debate at C-D. On its own, the debate is incomprehensible, since it refers to rules with which we are so far unfamiliar. An appendix at M. 2:3-4 has been added, therefore, detailing these rules, thereby enabling us to understand the debate.

Deut. 14:23 declares that produce consecrated as second tithe must be eaten in Jerusalem. M. 2:2 now asks whether or not having liquid absorbed through the skin is a form of consumption equivalent to that of eating, and therefore a permissible use of second tithe. This question emerges from the case of oil, since oil may be either eaten as a food or rubbed on the skin as a lotion. In A, Simeon reasons that anointing is not equivalent to eating. It follows, in his view, that the farmer may not anoint with oil declared second tithe.[11] Sages reason to the contrary that allowing the oil to be absorbed through the skin is comparable to ingesting it orally. They therefore permit the farmer to anoint himself with oil declared second tithe.

In C sages attempt to prove their point by adducing principles from the laws of heave-offering. They hold that what is true for produce consecrated in the status of heave-offering, which is eaten only by priests, is certainly true for produce consecrated in the status of second tithe, which is eaten by ordinary farmers. They point out that oil declared heave-offering may be used as a lotion. It follows, in their view, that oil declared second tithe may also be used as a lotion. Simeon, D, counteracts sages' argument by showing that what is true for produce declared heave-offering is not necessarily true for produce in the status of second tithe. To do so he turns to the laws operative in regard to fenugreek and vetches. Like oil, these plants may either be eaten or may be used as a non-food. Fenugreek is used as a shampoo and vetches serve as cattle fodder. Simeon states that when fenugreek and vetches are consecrated as heave-offering, they may, in fact, be used for either of their purposes. Yet, when they are in the status of second tithe, they may be used only as a human food. It follows, therefore, that the status of second tithe imposes upon produce in that status restrictions which the status of heave-offering does not. It follows, furthermore, that if these plants are restricted to use as a food when consecrated as second tithe, oil too should be restricted to use as a food when consecrated as second tithe.

We have taken for granted that vetches and fenugreek declared heave-offering may be used as a food or as a non-food, but that when they are in the status of second tithe they are used only as a food. The appendix at M. 2:3-4 supplies us with these facts. M. 2:3A first states that fenugreek declared second tithe enters the status of food as soon as the plant begins to sprout and must be eaten at that time. The notion that fenugreek in the status of second tithe is to be used as a food is the stringency referred to by Simeon in M. 2:2D. M. 2:3B-D take up the topic of fenugreek which has been declared heave-offering. Both Houses agree that such fenugreek may be either eaten or used as a shampoo. This is the fact invoked by Simeon in M. 2:2D. The dispute deals with the secondary issue of when fenugreek declared heave-offering is deemed to be a food, and so is susceptible to uncleanness, and when it is deemed a non-food, and so cannot be rendered unclean. The House of Shammai hold that since fenugreek may be eaten by humans, we must treat it as a food. That is, we must preserve it in a state of cleanness unless it is clearly being used for some other purpose. The Hillelites base their reasoning on the more usual use of the plant as cattle fodder. Since fenugreek is not normally used as a human food, they deem fenugreek to be insusceptible to uncleanness as food unless it is actually being prepared for human consumption.[12]

M. 2:4 now gives us the pertinent rules for vetches. We consider first vetches in the status of second tithe. A declares that such vetches are a human food, and so must be eaten, as was true of fenugreek in the status of second tithe. B qualifies A's rule, however, declaring that vetches in the status of second tithe are not treated as are consecrated foods. That is, unlike consecrated foods, vetches brought into Jerusalem may be taken out again (cf. M. 3:5-6).[13] M. 2:4C-E carry forward A's notion that vetches consecrated as second tithe are treated as food. At issue is what the farmer must do if the consecrated vetches become unclean. Such vetches cannot be used as second tithe. Tarfon suggests that the unclean vetches be broken into batches of less than an egg's bulk. This is the minimum volume which conveys uncleanness.[14] Since vetches in this form no longer convey uncleanness, they may be mixed in with other produce and consumed as second tithe. Sages (E), however, rule that the unclean vetches must be deconsecrated and clean foods then purchased in their stead. This is the normal procedure for disposing of foods in the status of second tithe which become unclean.[15]

M. 2:4F-J are concerned with vetches which are designated heave-offering. Again we assume that such vetches may be used either as a food or as a non-food.[16] As in M. 2:3B-D, the question before us is when the plant is considered a food, susceptible to uncleanness, and when the plant is deemed not to be a food, and thus insusceptible to uncleanness. M. presents us with four opinions. The extreme positions are held by Shammai and Aqiba, with Houses holding intermediate views. Shammai (I) holds that since the vetches may be eaten, we always deem them a food, and they must be preserved in cleanness. Aqiba (J), on the other hand, bases his opinion on the normal use of vetches. Since this plant is generally used as cattle fodder, he rules that the vetches are always insusceptible to uncleanness as food.[17] The Houses hold, however, that vetches must be treated according to the specific use the priest intends to make of them. Both Houses agree that feeding vetches to cattle removes them from consideration as a food. They also agree that soaking the vetches signals the priest's intention to use the vetches as food, and they now are susceptible to uncleanness. The Houses debate the implications of crushing the seed pods. The Shammaites interpret this procedure as being a step in preparing the plant for human consumption. The Hillelites take the opposing view and hold that crushed vetches are still not a food and are therefore insusceptible to uncleanness. The fact that the

Houses are made to debate with Shammai and Aqiba indicates that this unit is an artificial construction.

As we have seen, the Houses' dispute in M. 2:3-4 concerns the characteristic use of fenugreek and vetches, plants which may be used in a number of different ways. The issue is important because the plant's normal use determines how that plant is to be handled when it enters a consecrated status. The notion that consecrated produce is handled according to its normal use is precisely the principle adduced at M. 2:1A-C. The placement of M. 2:3-4 here, then, is a fine piece of redactional work. Not only do these two pericopae serve as an explanatory gloss to M. 2:2, but they also carry forward the theme introduced at the opening of the chapter.

A. Fenugreek [in the status of] heave-offering with which a priest's daughter shampooed her head[18] —

B. the daughter of an Israelite is not permitted (<u>ršyyt</u>) to shampoo [with the same fenugreek] after her [the priest's daughter],

C. but she rubs (<u>m^cgnt</u>) her hair against her [the priest's daughter's] hair [and acquires benefit from the fenugreek in this way].

T. 2:1a (p. 248, ll. 1-2
(T. Ter. 10:4)

Heave-offering is set aside for the exclusive use of priests and their dependents. Thus fenugreek in the status of heave-offering is used as a shampoo only by members of a priest's family. The point of B is that this restriction holds even after the priest's family has derived benefit from the plant. As long as the plant has use, it is reserved for the priests. Once the priest's daughter has soaped her head, however, she no longer has any use for the lather (C). This being the case, the Israelite woman may rub her head up against the soaped-up hair of the priest's daughter and so acquire some of the lather for herself.

D. <u>Fenugreek which is [declared to be] second tithe may be eaten [while it is still] in bud.</u>

E. <u>And [fenugreek] [declared to be] a heave-offering</u> —

F. "the House of Shammai says, 'Anything done with it is [done] in [a state of] cleanness,'

G. "and the House of Hillel says, 'Anything done with it is [done] in [a state of] uncleanness, except for shampooing with it' [= M. 2:3],"

H. the words of R. Meir.

I. R. Judah says,

J. "The House of Shammai says, 'Anything done with it is [done] in a [state of] cleanness, except for shampooing with it,'

K. and the House of Hillel says, 'Anything done with it is [done] in a [state of] uncleanness, except for soaking it' [= M. 2:3C-D]."

T. 2:1b (p. 248, l. 2 — p. 249, l. 6)
(y. M.S. 2:3)

Meir assigns to the Hillelites the opinion M. 2:3 ascribes to the Shammaites. Judah (I-K), however, agrees with the form of the dispute we have in M.

L. Vetches which are [in the status of] second tithe should be eaten [when] sprouting [M. 2:4A].

M. [As for vetches declared] heave-offering [M. 2:4F] —

N. The House of Shammai say, "They soak [them] in [a state of] cleanness, and they crush and feed [them to cattle] in [a state of] uncleanness [M. 2:4H],"

O. And the House of Hillel say, "They soak and crush [them] in [a state of] cleanness, and they feed [them to cattle] in [a state of] uncleanness [M. 2:4G]" [ed. princ. adds: the words of R. Judah].

P. R. Meir says, "The House of Shammai say, 'They soak and crush [them] in [a state of] cleanness, and they feed [them to cattle] in [a state of] uncleanness [M. 2:4G].'"

Q. And the House of Hillel say, "Anything done with them is [done] in [a state of] uncleanness [M. 2:4J]."

R. R. Yosé says, "This is the version (mšnt) of R. Aqiba.

S. "Therefore he ruled that they may be given to any priest [even an ᶜam ha'areṣ],[19] but (w-) the sages did not accept his opinion (l' hwdw lw)."

T. 2:1c (p. 249, ll. 6-10)
(Q-R: cf. M. Ḥallah 4:9)

In M. the Houses agree that vetches declared heave-offering are used either as a food or as cattle fodder. They disagree as to when it is to be deemed a food and thus susceptible to uncleanness. T. now introduces two alternate versions of the Houses' dispute. We deal first with Meir's version (P-Q) because his version differs most radically from the dispute as it appears in M. Although he does not change the view of the Shammaites, he assigns to the Hillelites the position held by Aqiba in M. 2:4J, namely that vetches declared heave-offering are never deemed a food. Judah, in N-O, disagrees with Meir's characterization of the Hillelites as Aqibans. He makes the Hillelites agree with the Shammaites that vetches declared heave-offering are sometimes deemed food and sometimes deemed cattle fodder. They disagree only on how crushing the seeds affects the plant's status. Yosé, in R, claims that Meir's version of the dispute is the one known to Aqiba. This is hardly surprising since, according to Meir, the Hillelites and Aqiba stand in agreement. Yosé goes on to use this opinion to explain another of Aqiba's rulings. Since Aqiba does not consider vetches in the status of heave-offering ever to be susceptible to uncleanness, he rules that they may be given even to a priest who is not careful about the laws of uncleanness.

2:5-6

A. Unconsecrated coins and coins [designated as] second tithe which [were mixed together] and [then] were scattered —
B. whatever he collects is collected as second tithe until he has restored [the value of the lost second tithe],
C. and the remainder [of what he collects] is unconsecrated.
D. If he mingled [consecrated and unconsecrated coins together] and scooped them up [by the handful] (ḥpn) —
E. [he deems what he has in each handful to be consecrated or unconsecrated] by proportion.
F. This is the rule: Those [items] which are collected [one by one] [are accounted first] as second tithe. Those [items] which are intermingled [and scooped up in bunches are deemed to be consecrated or unconsecrated] by proportion.

M. 2:5

A. A sela^{c} which is second tithe and [an] unconsecrated [sela^{c}] which were confused [such that the consecrated coin could not be identified] —

B he brings a sela^c's worth of [copper] coins and says,

C. "The sela^c which is second tithe, wherever it may be, is deconsecrated with these coins."

D. And [after consecrating the copper coins] he selects the finer [coin] between [the two sela^cs] (sbhn)

E. and deconsecrates [the copper coins] with it.

F. For they ruled, "They deconsecrate silver with copper [only] out of necessity. But [if they do deconsecrate silver with copper] it may not remain so, but they must immediately (ḥwzr w-) deconsecrate [the copper coins] with silver [coins]."

M. 2:6
(b. B.M. 55b-56a; D-E:
y. B.M. 4:5; cf. y. M.S.
2:7)

Consecrated coins have become confused with unconsecrated ones such that we can no longer distinguish which is which. We need to know whether the status of second tithe still adheres to those specific consecrated coins that are lost or whether we hold the status of consecration to be dispersed throughout the batch. The answer to this question determines how we go about replacing the lost consecrated money. If we hold the status of second tithe no longer is localized in specific coins, then any coins in the batch can be designated to replace the lost money (M. 2:5). If, on the other hand, the status of second tithe remains associated with the specific coins that were lost, these must be deconsecrated and their status transferred to other coins (M. 2:6). We will consider each of these proposed solutions in turn.[20]

According to M. 2:5, the farmer replaces the lost consecrated coins by randomly choosing from the mixture an amount of money equal in value to that which was lost. These coins are deemed to be consecrated in place of the original coins. For this solution to work, we must assume that the status of second tithe, which once inhered in the lost money now no longer is localized in specific coins. As a result, all of the coins in the mixture are of equal status, such that it is immaterial which specific coins the farmer draws from the batch to designate as second tithe. The conception is that the mixture of coins is analogous to a batch of untithed produce, which also has the consecrated status of tithes dispersed within it.[21] Just as the farmer randomly chooses an appropriate amount of produce from the batch to be heave-offering or tithes, so may he choose an appropriate value in coin from the mixture to replace the lost second tithe. This new batch is then deemed fully consecrated. This he does in either one of two ways, depending upon the circumstances of his retrieval. In A-C the farmer retrieves coins from the mixture one at a time. He designates each coin as second tithe until he has consecrated the appropriate amount of money he needs to replace what was lost. In the second case (D-E), the farmer scoops up a handful of coins from the pile. We assume that he has in his hands both some of the lost consecrated coins and some unconsecrated coins in the same proportion as these exist in the original mixture. This being the case, the farmer may randomly designate from what he has in his hand an appropriate proportion of coins to be second tithe. He continues to designate coins in this way until he has replaced the full value of the lost coins.

In M. 2:6 the farmer has two similar coins, one consecrated and one unconsecrated, and no longer knows which is which. Again the question is how the farmer recovers his consecrated money. The pericope proceeds under the assumption that sanctity still adheres to the individual coin which is lost. This is so, presumably, because the farmer had at one time designated that specific coin to be consecrated. As a result, the farmer may not simply pick any one of the coins and declare it to be second tithe. Rather, he must deal with the coin which is consecrated and which is now lost. This he does by first deconsecrating the lost coin, transferring its status to other coins. Now both coins in the batch are unsanctified. The farmer then selects the finer of these two coins and consecrates it as second tithe in place of the other coins.

The procedure outlined above provides the formulator of the law with an opportunity to make a second point in connection with the transfer of sanctity from one coin to another. Such a transfer should not take place from a coin of better metal (e.g., silver) to a coin of baser metal (e.g., copper) (F). The point is that the sanctity must always travel from items of less intrinsic worth to items of greater intrinsic worth.[22] The law here is based on an analogy with the sale of produce. When the farmer sells produce, he transfers the status of second tithe from items of little intrinsic value, fruit, to items of greater intrinsic worth, metal. Similarly, when he transfers consecration between coins, the transfer must take place from coins of baser metal to coins of finer metal.

A. Unconsecrated coins and coins dedicated[23] [to the Temple] which became scattered —
B. whatever he collects is collected as dedicated [coins] [cf. M. 2:5A-B],
C. whether it be coins [mixed with] coins, pieces of fruit [mixed with] pieces of fruit, or pomegranates with pomegranates, or anything which does [GRA, MB, HD, Lieberman add: not] intermingle (lybll).
D. But [if it is a substance] which does [delete: not] intermingle —
E. whatever he collects is collected as both [dedicated and unconsecrated in proportion],
F. and whatever is in excess, is in excess for both [dedicated and unconsecrated in proportion].

T. 2:4 (p. 249, l. 16 — p. 250, l. 19)

T. goes over the ground of M. 2:5-6, applying the rule of each pericope to a specific type of commodity. In this way, T. accounts for M.'s contradictory rules. We begin at A-C with discrete items, such as coins. Unconsecrated coins and coins dedicated to the Temple have become scattered together such that we no longer know which is which. As at M. 2:6, we hold that the status of dedication continues to adhere to the originally consecrated coins. Since we do not know which specific coins are dedicated, B rules that we must treat all the recovered coins as though they are dedicated to the Temple.

D considers the complimentary case. The intermingled items no longer retain their discrete identities. The result is an homogenous mixture, composed of a certain proportion of dedicated ingredients and a certain proportion of unconsecrated ingredients. We now follow the procedure described in M. 2:5. The farmer scoops up handfuls of the mixture and dedicates the appropriate portion out of each handful. He continues to do so until he has replaced what was lost. If the full value of the dedicated items has been recovered and some of the mixture still remains (E), we assume that the remainder has the same composition as the rest of the mixture had. The farmer continues to divide it into consecrated and unconsecrated batches by proportion until the mixture is used up.

A. A sela^{c} which is [in the status of] second tithe and [one] which is unconsecrated which were mixed together —

B. he brings a sela^{c}'s worth of [copper] coins [M. 2:6A-B].

C. Ben ^{c}Azzai says, "[He brings] two [sela^{c}s worth of copper coins]"

D. A sela^{c} which is unconsecrated and [one] which is dedicated [to the Temple] which were mixed together —

E. he selects the finer [coin] from them [the two sela^{c}s] and says [M. 2:6D],

F. "If [this] is dedicated, lo it is dedicated and if not, the dedicated [coin], wherever it may be, is deconsecrated with this [coin]."

G. And the second [sela^{c}] — lo, it is permitted [to be used as an unconsecrated coin].

T. 2:5 (p. 250, 11. 19-22)
(C: y. M.S. 2:7)

H. A sela^{c} which is [in the status of] second tithe and [one] which is dedicated [to the Temple] which were mixed together —

I. he brings a [sela^{c}'s] worth of [copper] coins and says,

J. "The sela^{c} which is second tithe, wherever it may be, is deconsecrated with these coins [M. 2:6C]."

K. He selects the finer of them [the two sela^{c}s] and says,

L. "If [this] is dedicated, lo it is dedicated, and if not, the dedicated [coin], wherever it may be, is deconsecrated with this [coin]."

M. He picks up the second [sela^{c}] and deconsecrates the [copper] coins with it.

T. 2:6 (p. 250, 11. 22-26)

N. A sela^{c} which is second tithe and [one] which is unconsecrated and [one] which is dedicated [to the Temple] which were mixed together —

O. he brings a sela^{c}'s worth of [copper] coins and says,

P. "The sela^{c} which is second tithe, wherever it may be, is deconsecrated with these coins."

Q. He selects the finest [coin] among them [the three sela^{c}s] [M. 2:6A-D] and says,

R. "If this [coin] is dedicated, lo, it is dedicated; and if not, the dedicated [coin] wherever it may be, is deconsecrated with this."

S. He [then] picks up the second [$sela^c$] and deconsecrates the [copper] coins with it.

T. And the third coin — lo, it is permitted [to be used as an unconsecrated coin].

U. For they ruled, "They deconsecrate copper [coins] with silver [coins], and silver [coins] with gold [coins] —

V. "silver [coins] with copper [coins] out of necessity — (M. 2:6F)

W. "but not gold [coins] with silver [coins]."

X. R. Eleazar b. R. Simeon says, "Just as they deconsecrate silver [coins] with copper [coins], so may they deconsecrate gold [coins] with silver [coins]."

Y. Rabbi said to him, "Why do they deconsecrate silver [coins] with copper [coins]?

Z. "Because they deconsecrate silver [coins] with gold [coins].

AA. "They may not deconsecrate gold [coins] with silver [coins] because they do not deconsecrate gold [coins] with copper [coins]."

BB. R. Eleazar b. R. Simeon says, "Gold [coins consecrated] as second tithe — they deconsecrate them with [copper] coins in Jerusalem."

CC. Under what circumstances?

DD. In the case of doubtfully tithed produce.

EE. But in the case of certainly untithed — lo, it is prohibited.

FF. In the case of doubtfully tithed produce, they even deconsecrate edibles with edibles.

GG. Gold coins [consecrated] as second tithe from doubtfully tithed produce which one made into utensils —

HH. it has no [means of] redemption.

II. If it was of silver — lo, this is permitted [*viz* to redeem the utensils made from the silver coins].

JJ. Second tithe from doubtfully tithed produce which became mixed together with [second tithe] from certainly untithed produce —

KK. Let them be consumed according to the [rule of the] more restrictive among them.

T. 2:7 (p. 250, l. 26 — p. 251, l. 39) (X-BB: y. M.S. 2:7; cf. M. Dem. 1:2)

These three Toseftan pericopae actually form one complex discussion centered on the proper handling of consecrated coins which are lost. Three smaller essays are discernible: A-T, U-AA, and BB-KK. I will discuss each of these essays separately.

The first essay, A-T, is made up of four subunits, each discussing a different combination of coins which have become mixed together. In each case, we assume that the sanctity adheres to specific coins. The purpose of this unit is to define the procedure by which the proper value of each status is restored. The first case, A-C, restates the situation presented in M. 2:6 — second tithe and unconsecrated coins. Ben ^{c}Azzai's gloss in C requires two sela^{c}s of copper coins in order to deconsecrate the second tithe, while M. only wanted one sela^{c}'s worth. Ben ^{c}Azzai reasons that since both sela^{c}s are in a state of doubtful sanctity, both coins must be deconsecrated. The second case is presented in D-G. Now an unconsecrated coin and a dedicated coin have been mixed together. Since the dedicated coin may be deconsecrated with any commodity, it is not necessary to provide copper coins. The farmer is able to rededicate the sela^{c} by simply choosing one of the silver coins and reciting the formula in F. The third case, at H-M, suggests a more complex problem. Now both coins fall into a status of sanctity, one being consecrated as second tithe and the other being dedicated. First the farmer must deconsecrate the coin sanctified as second tithe through the agency of copper, as required by M. We now have the situation of D-G — an unconsecrated coin and a dedicated coin mixed together. The procedure described in D-G is repeated. The finer coin is selected for the dedication, rather than being reserved for the consecration as second tithe. This is so because the dedicated coin enjoys a higher level of sanctity, being at the disposal of priests only and not of laymen, as is the case with second tithe. The remaining coin is now available to become second tithe in place of the copper coins.

The last unit of this essay, N-T, has coins of all three statuses mixed together. The procedure here is similar to those already described. First, the second tithe is deconsecrated. Then, the finest coin is selected and rededicated through the recitation of the standard formula. One of the remaining coins is consecrated as second tithe in place of the proper coins and the last coin is deemed to be unconsecrated.

Our second essay begins at U. The discussion centers around the kind of coins that can properly deconsecrate other coins which are imbued with the sanctity of second tithe. Two principles govern the formulation of this essay. There is general agreement that a coin may be deconsecrated by a coin stamped on a more valuable metal. In this case, sanctity is transferred to a commodity of greater desirability. The second datum in our essay is that silver coins may be deconsecrated with copper in cases of necessity. The debate between Eleazar and Rabbi centers around the propriety of other types of transactions in which a coin of baser metal is used to deconsecrate a coin of better material. Eleazar argues that just as silver is deconsecrated by copper, so may any coin be deconsecrated by one stamped on a baser metal. Rabbi argues that silver is an exception. That is, silver coins may be deconsecrated with copper coins, but gold coins, for example, may not be deconsecrated with silver. This is also the view of the Shammaites in M. 2:7.[24]

The final essay runs from BB-KK. This essay finds a place here because it continues Eleazar's rulings on the deconsecration of gold coins with copper coins. The theme of this essay, however, concerns the special consideration given to second tithe produce which is dema^{c}i, doubtfully tithed. If the produce is known to be untithed, then that which is removed in the name of second tithe is clearly second tithe. If, however, the produce is in a doubtful status, we take into consideration the fact that tithes may have already been removed. Now the material which was removed may not be second tithe at all. Consequently, the doubtfully removed second tithe is treated with greater leniency. Three cases of leniency are enunciated: BB, FF and II. In the first case, T. allows gold coins of doubtful status to be deconsecrated by copper coins, contrary to the principle that coins of lower value do not deconsecrate coins of higher value. FF allows produce designated as second tithe to be deconsecrated by other produce, even though comparable items usually do not deconsecrate each other. GG-II has the gold or silver coins used to deconsecrate the doubtful second tithe melted down and formed into a utensil. In the case of gold coins, the melting and recasting have changed the status of the gold. Since the gold is no longer considered a coin, it may not be deconsecrated with other coins, and the metal must remain consecrated. This is not true in the case of silver, however. According to MB, the silver utensil loses its status as coin but gains the status of produce. Since produce may be deconsecrated, even with other produce (as in FF), the silver utensil may be deconsecrated. Lieberman interprets this lemma in light of R. Dosa's ruling that second tithe produce may be acquired with a poorly minted coin ('symwn) (T. 1:4). The silver utensil, according to Lieberman's reading, may be considered a type of 'symwn and thus a valid coin in the view of R. Dosa.

JJ-KK return us to the issue which introduced this long discussion — coins of different statuses being mixed together. Through this device, the redactor signals the end of this long redactional unit. If the two types of second tithe have become mixed together, the entire mixture is treated as second tithe removed from certainly untithed produce. None of the leniency accorded to second tithe separated from dema^{c}i apply.

2:7

A. The House of Shammai say, "One should not convert (y^{c}śh) his [silver] sela^{c}s [consecrated as second tithe] into gold dinars."

B. And the House of Hillel permits [converting silver coins for gold coins].

C. Said R. ^{c}Aqiba, "I exchanged (^{c}śyty) silver [coins] for gold dinars for Rabban Gamaliel and R. Joshua."

M. 2:7 (A-B: b. B.M. 44b)

A farmer wants to reduce the number of consecrated coins he must take to Jerusalem by exchanging them for a lesser number of gold coins, which are of higher denomination. Gold coins rarely circulate, however, and so are not deemed to be a common medium of exchange. This being the case, the Shammaites prohibit the transaction. In their view, then, the status of second tithe should not be transferred to coins which rarely circulate and which the farmer might have trouble spending in Jerusalem (cf. M. 1:2). For the Hillelites, this practical consideration is irrelevant. Rather, they will allow any transaction in which the status of consecration is transferred from less desirable goods to more desirable ones. Since gold is a finer metal than silver, the transfer of second tithe from silver to gold is permissible in their view. Aqiba supports the Hillelite view, claiming he acts in accordance with it.

Q. One may sell his fellow a Tiberian tressis [in the status of second tithe] and receive from him a Sepphorite [tressis],
R. or [he may sell him a Sepphorite tressis and] receive from him a Tiberian (reading with E: ṭbryt) [tressis],
S. or [he may sell him] a dinar and receive from him two pundia,
T. or [he may sell him] a pundium and receive from him two issars.
U. In what [case] does this apply?
V. [In a case in which the coin was consecrated from produce separated] from doubtfully tithed produce.
W. But [in a case in which the coins were consecrated from produce separated] from certainly untithed produce,
X this [manner of transferring the status of second tithe from one coin to another] is forbidden.
Y. Also: in matters of doubtfully tithed produce they deconsecrate copper coins with [other] copper coins.
Z. One may sell his fellow a coin from Golan (reading with Lieberman: gwlnyt) [in the status of second tithe] and receive from him a dinar,
AA. or [he may sell him] a dinar and receive from him a pundium.
BB. Says R. Yosé, "In what [case] does this apply?
CC. In the case of [coins consecrated from produce separated as second tithe] from doubtfully tithed produce and in a case of necessity.
DD. but [in the case of coins consecrated from produce separated as second tithe] from certainly untithed produce, or as a matter of convenience (rywh),
EE. lo, this [manner of transferring the status of second tithe from one coin to another] is forbidden."

T. 4:13b (p. 265, l. 53 — p. 256, l. 57)

The pericope consists of two formally balanced units, Q-Y and Z-EE. S-T and Y expand the first unit. A farmer has coins in the status of second tithe and wants to transfer their status to other coins. Normally this is done only from coins of baser metal to coins of better metal (M. 2:6-9). T. rules that this restriction does not hold for coins consecrated by produce of doubtful status. Since these may not be consecrated at all, their disposition is treated leniently (T. 2:7BB-DD).[25] Three examples of such a transaction follow. In Q-R the transfer is made between coins of the same denomination, and therefore of the same metal. In S-T the status is transferred from a silver dinar to a copper pundia or from a copper pundium to copper issars. Finally, Z-AA has the farmer exchange one coin for coins of less total value. While such a transaction is generally permissible only for coins of doubtful status (W, Y, CC), CC declares that if there is a compelling need, it is also permissible for coins certainly in the status of second tithe (cf. M. 2:6).

A. [There is] a stringency [that applies] to dedicated items that does not [apply] to second tithe,
B. [and there is a stringency that applies] to second tithe that does not [apply] to dedicated items.
C. (1) Dedicated items are deconsecrated by means of any [other commodity],
D. [but] second tithe is deconsecrated by means of coins alone.
E. (2) Dedicated items do not increase [in level of sanctity] from then on [i.e., the time of dedication] ('yn ^{c}wlh m'yl; E: m'ylk; D: mk'n; ed. princ.: mk'n w'ylk),
F. [but] second tithe increases [in level of sanctity] from then on [i.e., the time of consecration] and not from then on [i.e., before consecration] (^{c}wlh m'yl wšl' m'yl).
G. (3) Dedicated items enter Jerusalem and are taken out [again],
H. [but] second tithe enters Jerusalem and is not taken out [again];
I. which is not the case for dedicated items.

T. 2:8 (p. 251, ll. 39-45)

J. [There is] a stringency [that applies] to dedicated items;
K. for (1) [the status] of dedication can apply to anything,
L. and (2) the laws of sacrilege (m^{c}ylh) apply to it,
M. and (3) it is permitted for consumption [by a non-priest] only if it has been deconsecrated,
N. which is not the case for second tithe.

T. 2:9 (p. 251, l. 45 — p. 252, l. 47)

T. supplies a lengthy discussion on the implications of various types of holiness. We compare rulings reflecting the sanctity of dedication to those reflecting the sanctity of second tithe. The comparison unfolds in two triplets, C-H and K-M, framed by balanced introductory and concluding formulae at B, I and J, N.

We turn first to the stringencies applied to second tithe holiness. Produce declared second tithe may be deconsecrated only with coins, while dedicated items may be deconsecrated with any commodity (C-D).[26] E-F is difficult. My translation follows Lieberman who relies on Maimonides (Yad M.S. 3:25). Produce declared to be second tithe may be dedicated to the Temple, thus entering a higher level of sanctity. This is possible because the produce is still in the farmer's possession and thus subject to his declaration. Dedicated items, however, may not pass to a higher level of sanctity (from maintenance of the Temple to dedication to the altar, for example). This is so because they are no longer in the farmer's possession and so he may not change their status. Finally, G-H notes that once second tithe is taken into Jerusalem, it must remain there until consumption. Dedicated items, on the other hand, may be taken into Jerusalem and then taken out again.

T. 2:9 turns to the stringencies applying to dedicated goods. Anything is subject to the status of dedication (K). In contrast, second tithe may be designated only from a limited range of goods, namely agricultural produce. Furthermore, the laws of sacrilege, the misuse of sanctified commodities, apply to dedicated items. This is because dedicated items are used by the priests (L). This type of transgression does not apply to second tithe. Finally, M. informs us that dedicated produce may be eaten by a non-priest only if it has first been deconsecrated. Second tithe, of course, is eaten by a non-priest, the farmer, in its sanctified state.

A. One who dedicates [to the Temple produce already in the status of] second tithe —

B. lo, he redeems it [i.e., sells it at such a price] so as to give (^{c}l mnt lytn) to the Temple its due ('t šlw) and [to set aside as] second tithe its due [i.e., the value of the produce originally consecrated as second tithe].

T. 5:22 (p. 272, ll. 64-65)

A farmer dedicates to the Temple produce already in the status of second tithe. He thereby commits the value of the produce for two conflicting purposes. T. declares that in such a case the farmer must sell the produce at twice its market value. In this way he is able both to give to the Temple the value he has pledged and at the same time, he is able to bring to Jerusalem the proper amount of money to be used to purchase produce for use as second tithe.

2:8-9

A. One who exchanges (*pwrṭ*) for a [silver] *sela*^{c} coins [sanctified as] second tithe —

B. the House of Shammai say, "The whole *sela*^{c} ['s worth of coins to be given in exchange] [must consist] of [copper] coins."

C. And the House of Hillel say, "[The *sela*^{c}*'s* worth of coins to be given in exchange may consist] of one *sheqel* [= half *sela*^{c}] of silver [coins] and one *sheqel* of [copper] coins."

D. R. Meir says, "They do not deconsecrate silver and produce with silver."

E. But (*w-*) the sages permit [the deconsecration of silver and produce with silver].

M. 2:8 (M. Ed. 1:9;
A-D: b. B.M. 45a)

F. One who exchanges a [silver] *sela*^{c} [sanctified as] second tithe [for other coins] in Jerusalem —

G. The House of Shammai say, "The whole *sela*^{c} [he receives must consist] of [copper] coins."

H. And the House of Hillel say, "[The *sela*^{c} he receives may consist] of one *sheqel* of silver [coins] and one *sheqel* of [copper] coins."

I. The disputants before the sages (*hdnym lpny ḥkmym*)[27] say, "[The *sela*^{c} may consist] of three silver *dinars* and [one] *dinar* of [copper] coins."

J. R. ^{c}Aqiba says, "[The *selac* may consist] of three silver *dinars* and a quarter (*wbrby*^{c}*yt*; some MSS: *wbrb*^{c}*t ksp wbrb*^{c}*t m*^{c}*wt*) [of the fourth *dinar* must consist of] [copper] coins."

K. R. Tarfon says, "[The fourth *dinar* may consist of] four *aspers* (*'spr*) of silver [equal to four-fifths of the *dinar's* value and the remaining asper must be of copper]."

L. Shammai says, "Let him deposit (ynyḥnh) it in a shop and consume its value [in produce] (wy'kl kngdh)."

M. 2:9 (M. Ed. 1:10;
A-C: b. B.M. 45a)

M. 2:8A-C and M. 2:9F-H form a doublet, with the Houses given the same opinions in each of the disputes. Only the superscriptions at A and F differ, thereby imposing different meanings on the two disputes. Each of the disputes is followed by supplementary material at M. 2:8D-E and M. 2:9I-K.

The pericope offers further analysis of problems regarding the transfer of sanctity from coins of different intrinsic value. A farmer has a batch of coins in the status of second tithe (A). He wants to exchange these coins for silver coins of higher denomination, thereby transferring the status of second tithe from the smaller coins to the silver sela^{c}. In this way, he reduces the number of coins he must take to Jerusalem. The Shammaites, B, rule that the batch to be given in exchange must consist entirely of coins stamped on copper. That is, they hold that each coin in the batch to be exchanged must be composed of a metal inferior in value to the silver coins he will receive. This is consistent with the principle found at M. 2:6F and 2:7 that outside of Jerusalem the status of second tithe is always transferred from coins stamped on baser metal to coins stamped on better metal. The Hillelites, C, allow a mixture containing both silver and copper to be exchanged for a silver sela^{c}. On the surface this ruling seems contrary to the principle that the status of second tithe is transferred only from coins stamped on baser metal to coins stamped on better metal. As we have already seen in M. 2:7, however, the Hillelites certainly agree that, outside Jerusalem, the status of second tithe passes from inferior metals to better metals. It seems, then, that in agreement with M. 2:5, they consider the status of second tithe to inhere not in the individual coins of a batch, but rather in the batch as a whole. A batch composed of both silver and copper coins, they hold, is of inferior material value on the average to one composed only of silver. They therefore allow the farmer to give a batch of silver and copper coins in exchange for silver sela^{c}s.

At D-E Meir and sages dispute a point similar to that of the Houses at B-C. As we shall see, Meir takes up the Shammaite opinion while sages adopt the Hillelite view. The question now is whether or not a mixture of silver and produce may be exchanged for a silver coin. Meir, D, rules that such a transaction is not allowed. Since produce in the status of second tithe may certainly be exchanged for silver coins, it appears that Meir's objection stems from the presence of silver in the collection to be exchanged. That is, Meir does not agree that the status of second tithe can be transferred from the silver coins in the collection to other silver coins. Meir, then, agrees with the Shammaites in B that the status of second tithe inheres in the individual items of a collection and not in the collection as a unit. Silver coins, even when mixed with produce, may not be given in exchange for other silver coins. Sages, however, do allow a collection of silver and produce to be exchanged for silver coins. They hold that the transfer of the status of second tithe takes place from the collection as a whole and not from its constituent elements.[28] In this way they agree with the Hillelites at C.

M. 2:9 repeats the dispute of M. 2:8 but with the transfer of the status of second tithe now taking place in Jerusalem. The farmer has a silver <u>sela^c</u> and wants to exchange it for coins of smaller denomination. Such coins are more easily spent in purchasing produce. As we shall see, the Houses are in agreement that inside Jerusalem the status of second tithe passes from coins stamped on finer metal to coins stamped on inferior metal, the reverse of what was true outside of Jerusalem. As before, however, they disagree as to whether the status of second tithe inheres in the individual coins or whether it inheres to the batch as a whole. The House of Shammai, G, rule that the farmer may receive only copper coins for his consecrated <u>sela^c</u>. As at B, they hold that the status of second tithe inheres in the individual coins and therefore each coin received in exchange for silver must be of the inferior metal. The Hillelites, H, rule that the farmer may receive a mixture of silver and copper coins for his silver <u>sela^c</u>. They hold, as before, that the status of second tithe inheres in the batch of coins as a whole and not in each of its component coins. Thus, as long as the metallic content of the batch as a whole is inferior to that of silver, the batch may be received in exchange for a silver <u>sela^c</u>.[29]

The authorities cited in I-K accept the Hillelite view that a collection of silver and copper coins is treated as a unit and therefore may be accepted in exchange for silver coins. Under discussion is what minimum amount of copper must be present in the batch in order for the batch as a whole to be deemed of lower quality than silver sela^c^s. The House of Hillel, in H, have indicated that at least half of the mixture must be of copper. This insures that the status of second tithe is transferred from silver to mostly copper coins. The three opinions which follow allow the presence of progressively smaller amounts of copper. The "disputants before the sages," I, require that only one dinar, one-fourth of a sela^c^, be of copper. Aqiba reduces the proportion of copper to one-fourth of a dinar, or one-sixteenth of a sela^c^. Tarfon reduces the proportion of copper even further, to one-fifth of a dinar, or one-twentieth of a sela^c^. I see no specific logic which stands behind the suggested proportions.

Shammai, L, is chronologically out of place, set in dispute with Yavneans. His opinion also stands radically apart on substantive grounds. He does not accept the premise which has generated the pericope before us, namely, that in Jerusalem the status of second tithe may be transferred from one coin to another. He argues that the silver sela^c^ may be used in Jerusalem only to purchase edible commodities. If the farmer does not want to purchase an entire sela^c^'s worth of produce at one time, Shammai rules that he must deposit the sela^c^ with a merchant and, over a period of time, purchase its value in produce to be eaten as second tithe.

A. R. Meir says, "They do not deconsecrate silver and produce with silver [= M. 2:8D],

B. "unless there is with it [i.e., the produce] [only] a small amount [of silver to bring the value of the produce up to a round figure]."

C. But the sages permit (the deconsecration of silver and produce with silver] [= M. 2:8E].

D. R. Jacob says in his [Meir's] name,

E. "[A silver sela^c^ is exchanged for] three dinars of silver and one dinar['s worth] of either copper or produce."

F. R. Simeon b. Eleazar says, "Shammai the Elder says, 'Let

<u>him deposit it in a shop and consume its value [in produce]</u> [= M. 2:9L].'"

T. 2:10 (p. 252, 11. 47-50)
(M. Ed. 1:9)

T. is composed of two autonomous disputes: A-C and D-F. The first qualifies Meir's opinion in M. 2:8D. He is now made to hold that there is a case in which produce and silver may in fact be exchanged for silver. The farmer has produce consecrated as second tithe which he wants to exchange for silver coins to be taken to Jerusalem. The produce, however, is worth slightly less than the silver coins. According to B, Meir allows the farmer to add to the produce some silver coins of small denomination consecrated as second tithe. In so doing, he creates a mixture of produce and silver which is of equivalent value to the silver coins he wants to take to Jerusalem. This mixture may now be given in exchange for the silver coins. By allowing silver and produce to be exchanged as a unit for silver, B brings Meir into essential agreement with sages, M. 2:8E, who also allow silver and produce to be exchanged for silver. In its context in T., then, the sages' opinion (C) loses its force. They now are made to permit what Meir, B, has already permitted.

D-F makes sense only as part of M. 2:9I-K's discussion. There M. is concerned with the kinds of coins a farmer may receive in Jerusalem in exchange for a silver <u>sela^c^</u> in the status of second tithe. In T., Meir rules that the farmer may receive up to three <u>dinars'</u> worth of silver in exchange for his consecrated <u>sela^c^</u>, with the fourth <u>dinar's</u> worth being exchanged for either copper coins or produce. That is, Meir holds that the status of second tithe may be transferred from silver to a mixture of silver and produce. His position here, then, is congruent to his view in A-C, and contrary to his opinion as we know it from M. Both A-B and D-E, then, radically revise Meir's position, setting him in essential agreement with the Hillelites in M. 2:8C and sages in M. 2:8E.

Simeon, F, cites Shammai's view, M. 2:9M, that coins in the status of second tithe are never exchanged in Jerusalem for other coins, but are used only for the purchase of produce.

2:10

A. One who has some dependents [in a state of] uncleanness and some [dependents in a state of] cleanness

B. lays down a [consecrated] _sela^c_ and says,

C. "Whatever the clean [dependents] drink, this _sela^c_ is deconsecrated with it."

D. It turns out that the clean [dependents] and the unclean [dependents] [may] drink from [the liquid contained in] the same jar.

M. 2:10

Whatever the farmer buys with consecrated money takes on the status of second tithe and must be eaten in Jerusalem in conditions of cleanness. In the pericope before us, the farmer purchases food some of which he intends to eat as second tithe and some of which he intends to treat as unconsecrated food by giving it to unclean persons to eat. According to M., the farmer must in this case stipulate that only that portion of the food which actually is eaten as second tithe is to be deemed in that status.[30] The rest of the food remains unconsecrated. In this way the farmer makes sure that the food his unclean dependents eat is not deemed to be consecrated and so protects himself against misusing consecrated food.

CHAPTER THREE
MAASER SHENI CHAPTER THREE

The bulk of Chapter Three regulates a farmer's use of consecrated coins or produce after he has brought them to Jerusalem. But for a brief ruling at M. 3:1, which serves as a transition between M. 2:10 and the present chapter, the rulings of Chapter Three fall into two major units. M. 3:2-4 investigate ambiguities regarding the purchase of food in Jerusalem with consecrated coins. The rest of the chapter, M. 3:5-13, is concerned with the effect of Jerusalem's holiness upon consecrated produce found within the city. At issue is whether the holiness of Jerusalem is analogous to that of the Temple's altar. If so, anything brought into Jerusalem for use as a tithe must be consumed within the city and may not be removed, just as sacrificial food, once placed on the altar, is sanctified and must remain on the altar. With this overview of the chapter's topics complete, let us consider each of these parts of the chapter in greater detail.

As I said, the first pericope of the chapter, M. 3:1, stands outside the larger concerns which follow. It carries forward instead the theme of M. 2:10, which permits the farmer to make his second tithe available to others to eat. M. 3:1 adds that he may do so only if he derives no personal benefit from the gift, for to do so would be to misappropriate that which is sanctified to a holy purpose. Thus a farmer may solicit his neighbor's aid in bringing second tithe to Jerusalem and, upon completion of the journey, share the second tithe with him. The farmer is forbidden, however, from offering part of the tithe as payment for the neighbor's labor. In the latter case, the farmer is purchasing a service with produce consecrated to God. Clearly, such a misappropriation is forbidden.

M. 3:1's concern for what a farmer may do with second tithe he has brought to Jerusalem sets the stage for the quite independent reflections of M. 3:2-4. At issue now is how the farmer uses coins in Jerusalem to consecrate food as second tithe. We recall from Deut. 14:24 that a farmer uses consecrated money in Jerusalem to buy food which he then eats as second tithe. The point of M. 3:2-4 is that the act of purchasing the food with consecrated money and the act of consecrating the food for use as second tithe are distinct from one another. M. 3:2 declares that it is possible to buy food with consecrated coins without in fact consecrating that produce for use as second tithe. The farmer, for example, might buy with money in the status of second tithe produce which is heave-offering. Although he buys the food, the anonymous opinion holds that he may not use it as second tithe. This is so because the food retains its status as heave-offering. It thus cannot be eaten by the farmer as second tithe even though it was bought with consecrated coins. M. 3:3-4 take up the complementary situation: cases in which food is consecrated without being purchased. This occurs when the farmer transfers the status of second tithe directly from his consecrated coins either to produce owned by a friend (M. 3:3), or to produce which he himself already owns (M. 3:4). In either case, neither the coins nor the food changes hands. Nonetheless, the coins become unconsecrated and the produce must be eaten as second tithe. The farmer thus has consecrated the produce without buying it.

The chapter's second large block of material (M. 3:5-13) considers the effect of Jerusalem itself upon consecrated produce brought within it. At issue is whether or not second tithe which is brought into Jerusalem is like sacrificial meat which is placed on the altar. M. Zeb. 9:1-2 declares that once meat appropriate to the altar is placed there, it becomes consecrated and may no longer be removed. The anonymous opinion in M. 3:5 applies the same logic to second tithe. Once this tithe enters Jerusalem, where it can be eaten, it may not be taken out of the city again (M. 3:5). M. 3:6 assumes this anonymous rule and investigates its applicability to food which will be separated as second tithe but which has not yet been so designated. The question is whether or not such food is already subject to the laws of second tithe. Having discussed this basic thesis, M.'s authorities consider problems connected with determining the city's boundaries in ambiguous cases. These are, for example, when fruit is on a tree limb which extends over the wall of the city or when the fruit is in a shed which straddles the city limits (M. 3:7-8). At issue is the legal location of such consecrated produce. If it is deemed to be in Jerusalem, the food may not be taken out of the city or sold, as M. 3:5-6 have stated.

M. 3:9 raises an important secondary issue regarding a possible exception to the law spelled out in M. 3:5-8. Now we have, inside Jerusalem, consecrated produce which is unclean. Unclean produce, as we recall, may not be eaten as second tithe (M. 2:10). Consequently, the farmer sells the unclean food and purchases other edibles in its stead. At issue is whether this unclean food, once deconsecrated, must remain in Jerusalem, in line with M. 3:5A-B, or whether it is to be taken out. The former view is taken by the Shammaites. Since the produce was in the status of second tithe when it entered Jerusalem, it must remain in the city even if it is later deconsecrated. The Hillelites hold, to the contrary, that once deconsecrated, the food is no longer subject to laws governing the use of tithes and so is taken out of the city.

The previous discussion assumes that unclean second tithe is sold even in Jerusalem so that it can be replaced with clean food. M. 3:10-13 carry this notion forward, applying it as well to cases in which the farmer purchases as second tithe items which he cannot eat as such. We deal first with produce (M. 3:10) or animals (M. 3:11) purchased in Jerusalem as second tithe. If this food is rendered unclean, it may be resold, just as is true of food originally designated as second tithe. M. 3:12-13 apply this same rule to inedible items purchased as second tithe. Since these cannot be eaten, they must be resold and replaced with food (cf. M. 1:5-6).

3:1

A. A man may not say to his friend,

B. "Take this produce [in the status of second tithe] up to Jerusalem [in order] to divide [it between us]."

C. But (_'l'_) he says to him [i.e., to the friend],

D. "Take this [produce] up [to Jerusalem] so that we may eat of it and drink of it [together] in Jerusalem."

E. Truly (_'bl_) they give [produce] to one another as a gift.

M. 3:1 (b. A.Z. 63a;
A-D: M. Sheb. 8:4;
T. Sheb. 6:25;
E: M. M.S. 1:1)

The gloss at E provides the exegesis for the contrasting case at A-B and C-D. While the text does not explicitly state that the produce under discussion is in the status of second tithe, the issue dealt with here, as well as the pericope's redactional context, require that this be the case.[1]

Since E is the key to the exegesis of the pericope, let us begin with its interpretation. As we know from M. 1:1, the only way a farmer may benefit from his produce designated as second tithe is by eating it in Jerusalem. E's point is that as long as such produce is consumed in Jerusalem without benefit to the farmer, he need not eat it personally. He may, therefore, give it as a gift to someone who will eat it under the proper conditions.

That this is in fact the point is brought out by the contrast between A-B and C-D. In A-B the farmer wants to enter into an agreement by which a friend will transport the farmer's consecrated produce to Jerusalem. From the language at B, it is clear that the friend is performing the work because he will receive some of the produce in return. Since, in this case, the farmer intends to use the consecrated food as payment for work, he will derive some benefit from his consecrated produce. It follows that such an arrangement is prohibited, as A states. In C-D, however, the farmer offers the produce to his friend as an unconditional gift, at the same time making a request that the friend help him get the food to the city. Since the farmer is not actually offering second tithe as payment for the friend's work, such an agreement is valid.

3:2

A. They do not purchase [produce in the status of] heave-offering with money [consecrated as] second tithe,

B. because it [the status of heave-offering] limits [the possibility of] its being used as a food (mm^c^ṭ b'kyltw) [since it can only be eaten by priests].

C. But (w-) R. Simeon permits [the purchase of produce in the status of heave-offering with money consecrated as second tithe].

D. Said to them R. Simeon, "If [mh omitted by most MSS] [the ruling] is lenient in regards to peace-offerings, which may be deemed to (mby'n lydy) [be impermissible for consumption under] the laws of refuse (pgwl), remnant (nwtr) and uncleanness (ṭm'), should we not rule leniently in regards to heave offering [which is not governed by these laws]?"

E. They said to him [i.e., R. Simeon], "If [the ruling] is lenient in regards to peace-offerings, which are permitted for non-priests [to eat], should we rule leniently for heave-offering which is forbidden for non-priests [to eat]?"

M. 3:2
(A-C: b. Zeb. 76a)

The dispute at A-C is followed by a debate at D-E. We expect this debate to explain the positions taken in the dispute. In fact, as we shall see, the debate introduces its own considerations. We have then two discussions of the same topic, one at A-C and one at D-E. Each presents a different reasoning for preventing a farmer from buying as second tithe produce in the status of heave-offering. Let us consider each discussion in turn.

A-B explain that a farmer may not purchase produce in the status of heave-offering for use as second tithe because he will not be able to eat such produce. The point is that the food consecrated as heave-offering will remain in that status and thus be prohibited to the farmer, even if it is bought with money in the status of second tithe. According to this view, then, food in one status of consecration cannot be converted to a different status, even if the owner wishes to change the status. Simeon, C, disagrees. He holds that any food purchased as second tithe takes on that status and can be eaten as such. Should the farmer purchase as second tithe produce in the status of heave-offering, in Simeon's view, that produce will lose its status as heave-offering and enter the status of second tithe. The farmer then will be permitted to eat that produce as second tithe. On this basis, Simeon can allow the farmer to purchase for use as second tithe food already consecrated as some other agricultural gift.[2] Thus, for Simeon, the farmer's intention to consecrate produce in a specific status is always effective, even if the produce is already in some other status of consecration.

D-F, however, ascribe to Simeon a far different reason for allowing the purchase of heave-offering for use as second tithe. He now agrees with A-B that the purchased food retains its status of heave-offering and thus is prohibited to the farmer. His point at D is that this is irrelevant. His reasoning is based on the already accepted laws of peace-offerings. These laws state that certain portions of an animal brought to the altar for use as a peace-offering are returned to the donor to be eaten by him. This meat may be consecrated as second tithe and eaten as such (M. 1:4). Simeon's point is that the meat may be purchased as second tithe even though it is liable in several ways to become entirely unfit for use as food. This occurs when the peace-offering is not eaten within two days after its slaughter (notar), when the priest who slaughters it intends to leave it uneaten for more than two days after slaughter (piggul) or when it comes into contact with a source of uncleanness.[3] From this Simeon concludes that it is permissible for the farmer to purchase as second tithe food which he might not be able to use. This is consistent with his view in C.

Simeon's argument is countered in E. Here sages reject Simeon's attempt to draw an analogy between peace-offerings and heave-offerings. Peace-offerings, to begin with, are permitted as food for both priests and non-priests. In this regard they are like any other food that the farmer might buy as second tithe. The fact that it might become unfit is deemed irrelevant. Produce in the status of heave-offering, to the contrary, is by definition permitted only to priests. On this basis, sages argue that what is true for peace-offerings, which to begin with anyone can eat, is not true for produce designated as heave-offering, which only priests can eat. Simeon's rationale for allowing priests to purchase heave-offerings for use as second tithe is thus shown to be without basis.

A. [If] he had two bins, one [for produce in the status] of heave-offering and one [for produce in the status] of second tithe,

B. and pieces of fruit between them —

C. [if the produce is found] nearer to the [bin for] heave-offering, it falls [under the laws governing] heave-offering,

D. [if the produce is found] nearer to the [bin for] second tithe, it falls [under the laws governing] second tithe,

E. [if it is found] halfway between [the two bins] they

apply the restrictive laws of both [the status of heave-offering and of second tithe].

F. R. Yosé b. R. Judah says, "A chest which they use for unconsecrated [coins] and [coins in the status of] second tithe,

G. and one found coins in it [and does not know the status of the coins],

H. if most [of the people] place consecrated [coins in the chest], [the coins are deemed to be] consecrated.

I. If most [of the people] place [coins in the status of] second tithe [in the chest, the coins are deemed to be in the status of] [Lieberman adds: second tithe].

J. R. Simeon said to them, "Lo, peace-offering, the breast and thigh of which are forbidden to non-priests [to eat]."

T. 2:11 (p. 252, ll. 50-55)
(A-E: cf. M. Sheb. 7:1;
F-I: b. Pes. 7a)

A-E and F-I are autonomous of M. They are placed here on account of E's notion that both the status of heave-offering and of second tithe can apply to a single item, as in the case proposed in M. 3:2C. In the first unit, A-E, we have two bins, one for produce in the status of heave-offering and one for produce in the status of second tithe. Pieces of fruit which someone meant to deposit in one of the bins are found on the ground. Since we do not know for which bin the produce was intended, we declare that the produce is in the status of the bin closest to where it was found. If the produce was found at a point equidistant from the two bins, it is treated according to the restrictions of both heave-offering and second tithe. That is, it must be eaten only by priests in a state of cleanness, like heave-offering, and it is forbidden to be eaten outside of Jerusalem or by a person in mourning, like second tithe. F-I makes the same point. Now coins are found in a chest which is sometimes used to contain unconsecrated coins and sometimes used to contain coins in the status of second tithe. Since we don't know the intention of the person who deposited the coins, we deem them to be in the status of the coins most generally deposited in the chest.

J continues the debate of the anonymous authorities and Simeon in M. 3:2D-E. Simeon argues in M. 3:2D that food designated as peace-offering is subject to more restrictions than is food in the status of second tithe, since peace-offerings are liable to the laws of refuse, remnant and uncleanness, which do not apply to second tithe. This argument is countered by sages (M. 3:2E) who point out that in fact heave-offering is more restricted in use since it may be eaten only by priests. Now, in T. 2:11J, Simeon is given the last word in the argument. He responds that there are parts of peace-offerings which, like heave-offerings, are also eaten only by priests. When an animal is offered as a peace-offering, its chest and thigh are consumed by the officiating priest. By this argument, he attempts to neutralize E by showing that the same restrictions applying to produce designated heave-offering also apply to animals declared peace offerings.

3:3-4

A. One who has coins [in the status of second tithe] in Jerusalem,

B. and needs them [for secular purposes],

C. and his friend has [unconsecrated] produce [in Jerusalem],

D. says to his friend,

E. "Lo, these coins are deconsecrated with your produce."

F. In consequence, this one [i.e., the friend] must eat his produce in [a state of] cleanness [because it is now in the status of second tithe],

G. but that one [i.e., the farmer] may use his coins for his needs [since they are no longer in the status of second tithe].

H. But (w-) he may not say this to an ^{c}am ha'areṣ, unless [the coins were] of doubtful status (dm'y).

M. 3:3 (H: cf. M. Dem. 1:2)

I. [One who has] [unconsecrated] produce in Jerusalem and coins [in the status of second tithe] in the provinces says,

J. "Lo, those coins are deconsecrated with this produce."

K. [One who has] coins [in the status of second tithe] in Jerusalem and [unconsecrated] produce in the provinces says,

L. "Lo, these coins are deconsecrated with that produce,"

M. provided that the produce is [subsequently] brought up to Jerusalem and eaten [there as second tithe].

M. 3:4 (Cf. y. M.S. 1:4)

These two pericopae deal with the transfer of the status of second tithe from coins to produce. We recall that Scripture allows a farmer to bring consecrated coins to Jerusalem and to use these to buy food, which then becomes second tithe. The point of the material before us is that the farmer can transfer the status of second tithe from money to produce without in fact buying the produce. That is, the status of second tithe can be transferred without the coins or the food changing hands. M.'s discussion unfolds in a series of three cases: A-H, I-J and K-M, each of which incorporates the same declaration on the part of the farmer (E, J, L).

We begin in A-H, with a farmer who arrives in Jerusalem with coins in the status of second tithe. He wants to deconsecrate these coins so that he can use them for secular purposes. According to E-G, he may do so simply by declaring the status of second tithe to be transferred from his consecrated coins to unconsecrated produce owned by a friend. The point here is that the farmer may now use the deconsecrated coins to purchase what he wishes without fear that the purchased item will be consecrated. His friend, on the other hand, keeps the food, but now must eat it as second tithe.

H's ruling is obvious. The person whose produce is consecrated must of course eat it in conditions of cleanness. It follows that the farmer should not transfer the status of second tithe to produce owned by an <u>ᶜam ha'areṣ</u>, that is, one who is not reliable in this matter. The law is lenient, however, if the coins' status as second tithe is in doubt.[4] Produce consecrated with this money is itself of uncertain status. We allow this produce to be eaten by an <u>ᶜam ha'areṣ</u> since, in all events, we cannot be sure the food is in fact consecrated.

I-J and K-M apply the procedure of A-H to a new situation. Now the farmer himself owns both the consecrated coins and the unconsecrated produce. The farmer therefore consecrates produce which he already owns. We consider two possible scenarios. In I-J the consecrated coins are outside of Jerusalem and the unconsecrated produce is in the city. K-M cite the complementary case. The consecrated coins are in Jerusalem and the unconsecrated produce is in the provinces. In both cases, the farmer may transfer the status of second tithe from the coins to the produce, even though there is no change of ownership involved.

M makes an important point in connection with the transaction of K-L. If the newly consecrated produce is located outside of Jerusalem, it must be taken to the city. This is in contrast to the laws applying to produce which was originally separated as second tithe, which may either be taken to Jerusalem or sold. The force of M, then, is to declare that produce bought as second tithe may not be sold again but must be taken to Jerusalem for consumption.

3:5-6

A. Coins [in the status of second tithe] enter Jerusalem and go out [i.e., after they have been brought into Jerusalem they may be taken out again],

B. but produce [which is in the status of second tithe] enters Jerusalem and does not go out [i.e., it must remain in the city until it is consumed].

C. Rabban Simeon b. Gamaliel says, "Also: produce [in the status of second tithe] enters Jerusalem and goes out [again]."

M. 3:5
(C: cf. M. Dem. 1:2)

A. [Untithed] produce, the processing [for use as food] of which is completed and which passed through Jerusalem —

B. [let produce separated from it as] second tithe be returned and eaten in Jerusalem.

C. [And as for produce] whose processing is not completed [and which passes through Jerusalem],

D. such as grapes [being brought] to the winepress or baskets of figs [being taken] to a drying place —

E. the House of Shammai say, "[Let produce separated from it as] second tithe be returned and eaten in Jerusalem."

F. And the House of Hillel say, "Let it be redeemed and [then] eaten in any place."

G. R. Simeon b. Judah says in the name of R. Yosé, "The House of Shammai and the House of Hillel did not disagree concerning produce the processing of which was not completed, that [produce separated from it as] second tithe may be redeemed and [then] eaten anywhere.

H. "About what did they disagree?

I. "About produce the processing of which was complete.

J. "For the House of Shammai say, 'Let [produce separated from it as] second tithe be eaten in Jerusalem' [= M. 3:5B].

K. "And the House of Hillel say, 'Let it be redeemed and eaten in any place'" [= M. 3:5C].

L. And [produce separated as second tithe from] doubtfully-tithed produce enters Jerusalem and comes out, and is [afterwards] redeemed.

M. 3:6

Before us is a complex essay dealing with the question of second tithe which has been brought into Jerusalem. The essay begins at A-B by stating that coins and produce are subject to different rules in this regard. Consecrated money which is brought into Jerusalem may be taken out again, while produce brought to the city must remain there. M. 3:6 assumes and amplifies B's rule in the name of the Houses. Simeon b. Gamaliel, M. 3:5C, rejects the whole notion that consecrated produce must remain in Jerusalem, and so stands outside the entire essay. With the general structure of the essay in mind, let us turn to a closer look at each of its parts.

We begin with the distinction in A-B between coins and produce. The point here is that the produce itself is deemed to be a consecrated offering. Bringing the produce into Jerusalem where it must be eaten, then, is like placing a sacrificial animal on the altar. We know from M. Zeb. 9:1 that an animal once placed on the altar may not be withdrawn. Similarly, B reasons, produce designated as second tithe which has been brought into the city may not be taken out. This consideration does not apply to coins since these are in all events not the offering itself, but only a temporary substitute which carries its status. Simeon's disagreement with B rests on a different understanding of second tithe. He does not see an analogy between second tithe and sacrificial animals in this regard. Rather, second tithe is simply a kind of produce to be eaten in Jerusalem. The mere fact that the produce has been brought into the city takes on no special significance in this view. Like the coins at A, it may be brought in and out at will, as long as it eventually serves its intended purpose and is ultimately eaten as second tithe in Jerusalem.[5]

M. 3:6 asks whether or not M. 3:5B's rule applies also to untithed produce which is in Jerusalem. We know that a portion of this produce will eventually be declared second tithe. Do we consider this portion of the produce, prior to its actual designation, to be already subject to the laws of second tithe so that it may not be taken out of the city? We have two versions of the Houses' dispute on this matter: A-F and G-K. In A-F the Houses agree that fully processed produce which has entered Jerusalem must be treated as if its second tithe had already been designated within it. If the produce subsequently is taken out of the city, what is later removed as second tithe is deemed to have absorbed the holiness of Jerusalem and so must be returned there. The reason is that since the produce was processed, and therefore suitable for tithing, it is made holy as tithe by entry into the holy city.[6] By analogy to the altar, Jerusalem makes holy whatever is suitable for tithing and which is intended for use there as tithe. The Houses disagree, however, as to whether or not the rule applies even to produce which is not yet processed (C-F) and therefore not yet suitable for tithing. The Shammaites hold that since what is ultimately separated as second tithe has already been in Jerusalem at one point, it must be returned there. The Hillelites disagree. Since the produce is unprocessed, it is not subject to the laws of consecration at all. The fact that the produce was in Jerusalem is of no concern. According to the Hillelites, then, Jerusalem renders holy only that produce which is suitable for sanctification, i.e., only produce which is subject to the laws of tithing.

Simeon has a different version of the dispute.[7] According to him, both Houses agree that unprocessed produce is entirely exempt from the law of M. 3:5B. This is so because the unprocessed produce, not being consecrated, does not become subject to laws governing the use of tithes. This is the Hillelite position in F. The disagreement concerns consecrated produce, that is produce which is ready to be eaten and thus liable to the removal of tithes. The Shammaites hold the same opinion as before, that this produce is subject to the laws governing the use of consecrated foods and so what is later separated as second tithe must be returned to Jerusalem. The Hillelites claim that since second tithe has not yet been designated in the batch, none of the produce is actually subject to the laws of second tithe. Since none of the produce was in the status of second tithe when it was first in Jerusalem, it follows that none of it needs to be returned.

L is independent of the foregoing, but is directly related to the issue of M. 3:5, the language of which it follows. It refers to the case in which second tithe is separated from produce which is dema'i, that is, produce from which tithes may already have been removed. Since it is possible that second tithe already was removed from the batch, that which the farmer now separates may not have the status of second tithe at all. We therefore do not deem the separated produce to be subject to the restrictions which apply to produce that is certainly in the status of second tithe. If produce separated from dema'i is brought into Jerusalem, it may be taken out of the city and sold.

K. R. Simeon b. Judah says in the name of R. Yosé,

L. "Thus said the House of Hillel to the House of Shammai, 'Do you not concede that [in the case of] produce the processing of which is not completed [and which was brought through Jerusalem] that (delete: if) produce is [separated] from it (šl' amended by Lieberman to šlh) as second tithe, it is redeemed [and] eaten in any place? [Also] produce the processing of which is complete [should be treated] in a similar fashion (kyṣ' bhn).'

M. "Said to them the House of Shammai, 'No. For you have said [in regards to] produce the processing of which was not completed, that one may renounce his ownership over it, so as to render it exempt from liability to heave-offering and tithes [which is not the case for produce whose processing is completed].'

N. "Said to them the House of Hillel, 'Even [in the case of] produce the processing of which is completed, one can separate produce as heave-offering and tithes [from it] for another batch.

O. 'And a further reason is that people are not liable to [the laws applicable to] heave-offering and tithes until [these gifts] are [actually] removed [cf. T. 3:11].'"

T. 2:11c (p. 252, l. 55 —
p. 253, l. 62; y. M.S. 3:6)

The unit before us supplements Simeon's version of the Houses' dispute in M. 3:6G-K. As we recall, a farmer brings through Jerusalem a batch of processed produce, that is, produce which is liable to the removal of heave-offering and tithes. The Shammaites hold that second tithe later separated from the batch must be returned to Jerusalem. The Hillelites hold, to the contrary, that the second tithe produce may be taken out of the city. In the debate before us, the Hillelites present an argument for their view. They point to the fact that second tithe separated from a batch of unprocessed produce which was in Jerusalem may be taken out (M. 3:6G). They now claim that the same law should apply as well to second tithe separated from processed produce. This is so, presumably, because in both cases the produce was not in the status of second tithe when it was in the city. It therefore should not be subject to the laws which govern the use of second tithe. The Shammaites respond, M, by claiming that processed produce is in fact already subject to the laws which govern the use of heave-offering and tithes. A farmer can exempt unprocessed produce from the removal of heave-offering and tithes, for example, simply by declaring the produce to be ownerless.[8] He may not, however, so exempt produce which has been processed for consumption, since such produce is already liable to the removal of these gifts. In N-O the Hillelites are given the final word in the debate. They point out that produce in one batch may be separated as heave-offering or tithe for produce in another batch.[9] Since the farmer does not need to separate tithes from each batch individually, it follows that we do not deem the laws of second tithe to apply to the produce of each particular batch. Rather, these laws apply only to produce which has been declared second tithe. The principle standing behind the Hillelite position is articulated explicitly at O. Here the Hillelites claim that the status of second tithe does not adhere to produce until the farmer actually designates certain produce to be second tithe.

As it appears in T., the position of Simeon's Hillelites is now made perfectly consistent with the rule of M. 3:5A-B. That is, the Hillelites now can agree that produce in the status of second tithe, once in Jerusalem, must remain there. They are made to disagree only about the secondary issue of whether or not this restriction applies as well to produce which has not yet been specifically designated as second tithe. T., then, has read M. as the redactor wants, <u>i.e.</u>, with M. 3:6 functioning as a secondary development of M. 3:5.

3:7-8

A. A tree which is standing inside [Jerusalem] and [a bough of which] extends outside [the city],

B. or [which] is standing outside and [a bough of which] extends inside [Jerusalem] —

C. that which is above [the center of] the wall and inwards is [deemed to be] within [Jerusalem] [and that which is over the center of] the wall and outward is [deemed to be] outside.

D. Buildings containing olive-presses, the entrances of which are inside [Jerusalem] and the contained spaces of which are outside,

E. or the entrances of which are outside and the contained spaces of which are inside —

F. the House of Shammai say, "It is all [deemed to be} inside,"

G. and the House of Hillel say, "That which is opposite [the center of] the wall and inward is [deemed to be] within [and that which is opposite the center of] the wall and outward is [deemed to be] outside."

M. 3:7 (A-C: b. Mak. 12a,
D+G: y. M.S. 3:7)

H. the [Temple] chambers built in the holy [precinct] and open to the unsanctified [area] —

I. their inner space (<u>twkn</u>) is [deemed to be] unsanctified and their roofs are [deemed to be] sanctified.

J. [Those] built in the unsanctified [area] and open to the sanctified [precinct] —

K. their inner space is [deemed to be] sanctified and their roofs are [deemed to be] unsanctified.

L. [Those] built [partly] in the holy [precinct] and [partly] in the unsanctified [area] and open to [both] the sanctified [precinct] and the unsanctified [area] —

M. [as for] their inner spaces and their roofs —

N. [that part which is] in the sanctified [precinct] and inward (<u>mkngd hqdš wlqdš</u>) is [deemed to be] sanctified [and that part which is] in the unsanctified [area] and outward is [deemed to be] unsanctified.

M. 3:8 (H-I: b. Pes. 86a; J-K: b. Zeb. 56a; cf. T. Kel. B.Q. 1:11)

These pericopae are concerned with areas which straddle the boundaries of sacred precincts and which therefore are of ambiguous status. The question is whether or not such areas are deemed to belong entirely to the sanctified domain. The material before us has been formulated independently of the laws of second tithe. It is placed here because of our chapter's concern with laws which apply only within the bounds of Jerusalem. Formally we have a triplet, A-C, D-G and H-N, whose units are mildly apocopated.

A tree has a bough hanging over the wall of Jerusalem (A). C rules that only that part of the limb which is actually located within the walls of Jerusalem is deemed to be in the city. Thus, for example, if a farmer picks fruit from that part of the limb which is inside the city, the fruit is deemed to have been in Jerusalem, and if it is declared second tithe, must remain in the city.[10] The converse is true if the fruit is picked from part of the limb which is outside the wall. Our sole consideration then is the actual location of each part of the limb in relation to the city walls.[11]

In D-G we consider a shed which intersects the walls of Jerusalem. The Hillelites rule that, like the tree limb in A-C, only that part of the shed which is actually located within the walls is deemed to be in the city. The Shammaites, however, claim that the entire contained space of the shed is deemed to be like Jerusalem itself. They hold that the outer wall of the shed establishes a new boundary-line for Jerusalem. The Shammaites, apparently, would hold that any second tithe produce anywhere in the shed may be eaten. In this they reject the Hillelite notion, expressed at G, that the domain of Jerusalem is restricted to its traditional boundaries.

H-N turns our attention to chambers built partially in the Temple courtyard and partially outside the courtyard. Here we are told that the status of the roof and the status of the contained space are judged according to different criteria. The rule concerning the status of the roof follows the Hillelite position that the domain of the sacred does not extend beyond its traditional bounds. It follows that the roof will always have the status of the ground over which it is built (H-K). If the roof straddles the courtyard wall, each segment of the roof will have the status appropriate to its actual location (L-N). The inner space of the chamber, however, is deemed to be part of the domain to which it is contiguous. That is, like the Shammaite position above, the outer walls of the chamber are deemed to establish new boundary lines for the area into which the chamber opens. If the entrance of the chamber is located in the courtyard, for example, the walls of the chamber which extend beyond the traditional courtyard wall become the new boundaries of the courtyard and the chamber's entire contained space becomes part of the courtyard's domain. This principle, however, does not help us determine the status of the chamber if it opens out to both the courtyard and the unsanctified area (L-N). In such a case, we follow the Hillelite position and deem each part of the chamber's contained space to have the status appropriate to its actual location in relation to the traditional boundary line. H-N, then, offers a compromise between the position of the two Houses.

A. A tree which is standing inside [Jerusalem] and a bough of which extends outside [M. 3:7A] —

B. [produce picked from it and designated as] [second] tithe has [the possibility of] redemption [i.e., it may be given in exchange for coins].

C. If the branch came back inside [Jerusalem], [produce picked from it and designated as] second tithe does not have [the possibility of] redemption.

D. [If] it was standing outside and its bough extends inside [M. 3:7B] —

E. [produce picked from it and designated as] second tithe does not have [the possibility of] redemption.

F. If the branch came back outside, [produce picked from it and designated as] second tithe has [the possibility of] redemption.

G. Buildings containing olive-presses (bty btym: reading with E: bty bdyn) the entrances of which are inside [Jerusalem] and the contained spaces of which are outside [Jerusalem],

H. or the entrances of which are outside and the contained

spaces of which are inside [M. 3:7D-E] —

I. the House of Shammai say, "They do not redeem [produce in the status of second tithe] inside them, as if they were [located totally] within Jerusalem, and they do not eat Lesser Holy Things as if they were [totally] outside,

J. and the House of Hillel say, "That which is opposite [the center of] the wall and inward is [deemed to be] within [and that which is opposite the center of] the wall and outward is [deemed to be] outside [M. 3:7G]."

K. Said R. Yosé, "This is the version of R. Aqiba.

L. "The original version (mšnt r'šnh) [is as follows]:

M. "The House of Shammai say, 'They do not redeem [produce in the status of] second tithe in it, as if they were inside [Jerusalem], and they do not eat Lesser Holy Things in them, as if they were outside.'

N. "and the House of Hillel say, 'Lo, they [are treated] like [Temple] chambers; the one whose entrance opens to the inside [is deemed to be] inside and the one whose entrance opens to the outside [is deemed to be] outside'" [cf. M. 3:8H-L].

T. 2:12 (p. 253, l. 62 — p. 254, l. 72; G-N: T. Arak. 5:15)

O. A [Temple] chamber built in the holy [precinct] with (w) its entrance [open] to the unsanctified [area] —

P. its inner space is unsanctified and its roof is sanctified [M. 3:8H-I].

Q. They do [Lieberman adds with E: not] eat Most Holy Things in it, and they do not slaughter Lesser Holy Things in it, and they are not responsible on its account [for the laws of] uncleanness.

T. 2:13 (p. 254, ll. 72-74)

R. [A chamber] built in the unsanctified [area] with its entrance [open] to the holy [precinct] [M. 3:8J],

S. even though it should not remain this way —

T. its inner space is sanctified and its roof is unsanctified [M. 3:8K].

U. They eat Most Holy Things in it (Lieberman adds with D: and they slaughter Lesser Holy Things in it) and they are liable on its account for [the laws of] uncleanness.

T. 2:14 (p. 254, ll. 74-77)
(R-T: b. Yoma 25a, Zeb. 56a)

V. [A chamber] built in the holy [precinct] and open to the unsanctified and the sanctified [areas] —

W. it is all [deemed to be] sanctified.

X. [A chamber] built in the unsanctified [area] and open to the sanctified and unsanctified [areas] —

Y. it is all [deemed to be] unsanctified.

Z. [A chamber] built in the sanctified and unsanctified [areas] and open to the sanctified (Lieberman adds: and unsanctified) [areas] [M. 3:8L] —

AA. [as in the case of] the chamber of fire (lškt byt hmwqd).

BB. [As for] their inner space and their roofs —

CC. [that which is] above the sanctified [area] (mkngd hqwdš wlqwdš) is [deemed to be] sanctified

DD. [and that which is] above the unsanctified [area] is [deemed to be] unsanctified [M. 3:8M-N].

EE. If [the case is one] of eating holy things (E: matters of uncleanness), [the status of] the entire [chamber] is according to [the location of] its entrance.

FF. If [the case is one] of matters of uncleanness (E: eating holy things), [the chamber is deemed to be sanctified only] from the [center of the] wall and inwards (Lieberman deletes with E: it is all according to the [location of] its entrance).

GG. [If a chamber] was built on the wall (Lieberman: of the [Temple] courtyard),

HH. they eat Most Holy Things in it and they (Lieberman deletes with E: do not) slaughter Lesser Holy Things in it,

II. [If a house] was built on the wall (Lieberman: of Jerusalem),

JJ. they (Lieberman: eat) Lesser Holy Things in it but they do not redeem [produce in the status of] second tithe in it.

KK. A house built on the wall [of a walled city, Lev. 25:29] —

LL. R. Judah says, "It is [treated] as if it were outside [the walls]."

MM. R. Simeon says, "It is [treated] as if it were inside [the walls]."

NN. All of the [Temple] chambers are built totally in the sanctified [precinct].

OO. R. Yosé says, "The border lines of the Land of Israel that are [given] in Scripture are judged to be outside of the Land [of Israel]."

PP. R. Eleazar b. R. Yosé, "They are judged to be in the Land of Israel."

T. 2:15 (p. 254, l. 77 — p. 255, l. 86)
(V-Y: y. M.S. 3:7; NN-OO: cf. T. B.Q. 8:19); KK-MM: T. Arak. 5:14)

The pericopae before us constitute an expansion of the material introduced in M. 3:7-8. It may be divided into four sections, corresponding to the thematic units of M. A-F cite and gloss M. 3:7A-C; G-N cite and gloss M. 3:7D-G; O-FF cite and gloss M. 3:8; and GG-PP are independent of M., although related in theme.

A-F expand M.'s discussion of the tree with the extended bough. C and F add the case of a branch which extends into the second area and then bends back to the first. Fruit on each segment of limb is still judged according to its actual location inside or outside the wall. G-N turns our attention to the olive-press which straddles the city wall of Jerusalem. The question is the status of the shed's contained space. Now we have two versions of the Houses' dispute. The first version, attributed to Aqiba (K), repeats the Hillelite position of M. 3:7G, but slightly alters the Shammaite view. The Hillelites, as before, hold that the inner space of the shed is divided according to actual location in relation to the city wall. The Shammaites now hold that the interior space is treated as a homogenous unit and that the stringencies of both locations apply throughout. R. Yosé gives a slightly different version of the dispute (K-N). He makes the Hillelites agree with the Shammaites that the entire inner space is deemed to have one status. The disagreement now is only about what that status is. The Shammaite position is as above, that the stringencies of both locations apply to the space. The Hillelites treat the sheds like the Temple chambers, whose entire inner space is given the status of the area contiguous to the entrance. Redactionally, this Hillelite opinion forms a bridge from the theme of G-N, olive-presses, to that of O-FF, Temple chambers.

Our third unit, O-FF, expands M.'s laws dealing with Temple chambers which are built across the boundary between the Temple courtyard and the unconsecrated ground. T.'s glosses at Q and U simply reiterate the law. Most Holy Things are eaten, Lesser Holy Things are slaughtered, and the laws of uncleanness apply only in the sanctified precincts of the Temple. Q and U have these same laws apply to those inner spaces which are deemed sanctified. V-Y add a further possibility for the location of the chamber besides those enumerated by M. Now the chamber is in one area while the entrance is open to both sanctified and unsanctified precincts. In this case, the entire chamber is given the status of the ground upon which it is built. EE and FF qualify the preceeding rule by claiming that we take into account the activity to be carried out in the chamber. If a priest wishes to eat Most Holy Things within the chamber, the status of the inner space is judged according to the location of the entrance. If we are concerned with the laws of uncleanness, only those parts of the chamber which are actually inside the courtyard wall are deemed to be in the status of the courtyard.[12]

GG-PP, as we have said, supply material autonomous of M. Before us is a series of cases in which an identified location sits directly over a boundary line. In each of the cases, the area on top of the boundary is deemed to take on the status of the more sanctified area. Thus the Temple chamber at GG-HH is deemed to be within the courtyard and sacrifices may be slaughtered and eaten there. The house in II-JJ is deemed to be within Jerusalem. Lesser Holy Things may be eaten there, and produce in the status of second tithe may not be given in exchange for coins in the house.[13] KK-MM concerns houses built on the walls of a city. The law referred to here states that if houses within a walled city are sold, they may be redeemed by the seller during the first year. After that time, they belong to the buyer in perpetuity. Houses outside the wall, in contrast, may be redeemed at any time, and if they are not redeemed, they nonetheless return to the seller during the Jubilee. Simeon, MM, holds that houses built on the city wall are deemed to be within the city, consistent with the principle of GG-JJ.[14]

NN refers us to Temple chambers. These, we are told, are never built on the courtyard wall, but only clearly within the courtyard. Lieberman relates this lemma to the gloss at T. 2:14S which declares that chambers built on unsanctified ground and opening onto the courtyard should not be allowed to remain.

OO-PP concerns geographical locations mentioned as the boundaries of the Land of Israel. Leazar's opinion, that the geographical points are within the Land of Israel, is consistent with the notion that areas directly on top of a boundary are imputed the status of the more sanctified area.

3:9

A. [Produce in the status of] second tithe which entered Jerusalem and was rendered unclean,
B. whether it was rendered unclean by a Father of uncleanness or whether it was rendered unclean by an Offspring of uncleanness,
C. whether [it was rendered unclean] inside [Jerusalem] or outside [Jerusalem] —
D. the House of Shammai say, "Let it all be redeemed and eaten inside [Jerusalem],
E. "except for that which was rendered unclean by a Father of uncleanness outside [Jerusalem], [which must be taken out]."
F. And the House of Hillel say, "Let it all be redeemed and eaten outside [of Jerusalem],
G. "except for that which was rendered unclean by an Offspring of uncleanness inside [of Jerusalem], [which may remain in the city]."

M. 3:9 (Cf. y. M.S. 3:8;
M. Sheq. 8:6-7)

Produce in the status of second tithe which becomes unclean may not be eaten as second tithe (M. 2:10). Rather, it must be deconsecrated so that other food can be purchased and eaten as second tithe in its stead. At issue now is whether unclean produce which is no longer usable as second tithe must remain in Jerusalem even though it is deconsecrated, in accordance with the principle of M. 3:5A-B, or whether the produce, once deconsecrated, may be taken out of the city. As we shall see, the Houses in fact take up positions between these two extremes, claiming that in some cases the produce must remain in the city, and in other cases, it must be taken out.[15] Formally, we have a superscription, A+B-C, introducing a Houses' dispute, D-G, in which the opinions are perfectly balanced opposites.[16]

In D-E the Shammaites rule that unclean produce in the status of second tithe remains in Jerusalem unless it is rendered unclean outside the city by a Father of uncleanness.[17]

That is, they hold that in the majority of cases the produce is treated like all other second tithe, in accordance with M. 3:5 A-B. Once it is brought into the city, it must remain there, even if it is subsequently deconsecrated. They make an exception, however, if the produce enters Jerusalem in a state of virulent uncleanness, the result of contact with a Father of uncleanness. Only in this extreme case do the Shammaites hold that the produce must be taken out of the city again. The Hillelites show more concern for the purity of Jerusalem. They hold that unclean produce, once deconsecrated, should be removed from the holy city (M. 3:5C). They make an exception only for produce tainted with minor contamination after it has already been brought inside the walls of Jerusalem. Since the uncleanness is only minor, they do not require the farmer to take the trouble to cart this produce away. We can, then, summarize the Houses' positions as follows: Both Houses agree on the extreme uses: produce made virulently unclean outside Jerusalem must not remain in the city, while produce made only slightly unclean after being in the city does remain there. As regards the remaining possibilities, produce made highly unclean in Jerusalem or slightly unclean outside the city, the Shammaites require the produce to remain in the city, in line with M. 3:5A-B, while the Hillelites rule that the unclean food is to be taken out.[18]

A. "[Produce in the status of] second tithe which entered Jerusalem and was rendered unclean,

B. "whether it was rendered unclean by a Father of uncleanness or whether it was rendered unclean by an Offspring of uncleanness,

C. "whether inside [Jerusalem] or outside [Jerusalem] [M. 3:9A-C] —

D. "the House of Shammai say, 'Let it all be redeemed and eaten inside [Jerusalem]';

E. "and the House of Hillel say, 'Let it all be redeemed and eaten inside [Jerusalem], except for that which was rendered unclean by a Father of uncleanness outside [Jerusalem],'"

F. the words of R. Meir.

G. R. Judah says, "The House of Shammai say, 'Let it all be redeemed and eaten inside [Jerusalem] except for that which was rendered unclean by a Father of uncleanness outside [Jerusalem],'[19]

H. "and the House of Hillel say, 'Let it all be redeemed and eaten outside [Jerusalem], except for that which was rendered unclean by an Offspring of uncleanness inside [Jerusalem] [M. 3:9D-G].'"

I. R. Eleazar says, "[If] it is rendered unclean by a Father of uncleanness, whether inside or outside [Jerusalem], let it be redeemed and eaten outside [Jerusalem]."

J. "[If] it is rendered unclean by an Offspring of uncleanness, whether inside or outside, let it be redeemed and eaten inside."

K. R. Aqiba says, "[If] it is rendered unclean outside [Jerusalem], whether by a Father of uncleanness or by an Offspring of uncleanness, let it be redeemed and eaten outside [Jerusalem].

L. "[If] it is rendered unclean inside [Jerusalem], whether by a Father of uncleanness or by an Offspring of uncleanness,[20] let it be redeemed and eaten inside [Jerusalem]."

M. Said R. Simeon b. Eleazar, "The House of Shammai and the House of Hillel did not disagree about [produce] that was rendered unclean by a Father of uncleanness outside [Jerusalem], that it may be redeemed and eaten outside, nor about that which was rendered unclean by an Offspring [of uncleanness] inside [Jerusalem], that it should be redeemed and eaten inside [Jerusalem].

N. "About what did they disagree?

O. "About that which was rendered unclean by a Father of uncleanness inside [Jerusalem] and [about that which was rendered unclean by] an Offspring outside [Jerusalem].

P. "For the House of Shammai say, 'Let it be redeemed in the Place [i.e., Jerusalem] and eaten in the Place,'

Q. "And the House of Hillel say, 'Let it be redeemed in the Place and eaten in any place (bkl hmqwmwt).'"

T. 2:16 (p. 255, l. 86 — p. 257, l. 101; I: cf. M. Sheq. 8:7)

T. cites M. 3:9 and offers alternative apodoses for the dispute. I discern in T. three units: A-H, I-L and M-Q. In A-H Meir and Judah dispute over the proper version of the Houses' opinions. Judah's version is the same as M. 3:9D-G. Both Houses agree that in some cases consecrated produce which is rendered unclean is taken out of Jerusalem. Meir's version of the dispute, however, has the Shammaites hold that in all cases the produce must remain in the city. This is in line with the view of the autonomous authorities in M. 3:5A-B. We see, then, that while Judah holds that the Houses' dispute is irrelevant to M. 3:5's dispute, Meir holds that the Houses clearly stand on the side of the anonymous rule.

An entirely different version of the dispute, attributed to Yavneans, appears in I-L. Here the authorities do not know the concerns of M. 3:5, but focus, rather, on how the produce became unclean in the first place. Eleazar rules that any produce made unclean by contact with a Father of uncleanness, and which is therefore in a state of virulent uncleanness, must be taken out of the city. Only produce mildly unclean stays in. Eleazar, then, has a definite interest in keeping out of the holy city produce that is severely contaminated.[21] Aqiba makes a distinction on the basis of the location of the produce when it becomes unclean. If it becomes unfit for use as second tithe outside Jerusalem, it may be taken out again, since it was not capable of being used as second tithe already at the time it entered the city. Produce fit for use as second tithe when it enters the city must remain there, however, even if it is later rendered unclean.[22]

Simeon b. Eleazar's description of the dispute (M-Q) is congruent to the Houses' opinions as we know them from M. 3:9 D-G. He has simply specified the points of agreement (M) and disagreement (O-Q).

3:10

A. [Produce] purchased with coins [in the status] of second tithe, which becomes unclean [and therefore may not be eaten as second tithe] —

B. let it be redeemed.

C. R. Judah says, "Let it be buried."

D. They [the authorities of A-B] said to R. Judah, "If [it is the case that when produce which is designated as] second tithe itself becomes unclean, lo, it must be redeemed, is it not logical that produce purchased with coins [in the status of] second tithe which becomes unclean [also] should be redeemed?"

E. He said to them, "No! If you say this in regard to [produce designated as] second tithe itself, which, if in [a state of] cleanness, may be redeemed when it is outside [Jerusalem] (brḥwq mqwm), can you say so as regards produce purchased with coins [in the status of] second tithe which, when it is [in a state of] cleanness, may not be redeemed when outside [Jerusalem]?"

M. 3:10 (y. Pes. 2:4;
A-C: b. Pes. 38a; b. Zeb. 49a)

As we have seen in M. 3:9, produce designated second tithe which becomes unclean may be deconsecrated even after it has been in Jerusalem. M. now turns our attention to produce which acquired the status of second tithe by being purchased with coins in that status. Since such produce is purchased in Jerusalem, it normally may not be taken out of the city and deconsecrated (M. 3:5A-B). What happens if such produce becomes unclean, however, and may no longer serve as second tithe? The anonymous authorities and Judah dispute whether or not in this situation the farmer may, in fact, deconsecrate it, purchasing other produce in its stead. Their dispute is followed by a debate in D-E.

A-B hold that food bought with coins in the status of second tithe and which becomes unclean may be deconsecrated. The reasoning is given in D. Produce which has been separated as second tithe may be deconsecrated if it becomes unclean (M. 3:9). It follows that produce purchased with coins in the status of second tithe may also be deconsecrated if it becomes unclean. That is, produce purchased as second tithe is treated in the same way as is produce originally separated as second tithe. Judah, E, rejects this reasoning. He points out that produce originally separated as second tithe may be sold outside Jerusalem. This is not true, however, for produce purchased as second tithe. It follows, Judah claims, that while the farmer does have the option of deconsecrating produce originally designated as second tithe, he does not have this option in regard to produce purchased as second tithe. If such produce becomes unclean, it follows, the farmer has no choice but to leave it to rot.

A. [Produce] purchased with coins [in the status] of second tithe which becomes unclean,

B. let it be redeemed.

C. R. Judah says, "Let it be buried."

D. They said to R. Judah, "Should you rule more strictly [in regards to] what is secondary (bṭpylh) than [in regards to] what is primary ($b^{c}qr$) [M. 3:10A-C]?"

E. He said to them, "We find [cases in which] they rule more strictly in regards to what is secondary than [in regards to] what is primary. For [in the case of] a substitute (tmwrh), its [status of] dedication applies to it [even if] it is afflicted with a permanent blemish, but [the status of] dedication [for use on the altar] does not apply (delete: except) to [the animal the farmer originally proposes to dedicate] if it is afflicted with a permanent blemish."

F. They said to him, "[Our case is proven] from the example you bring (mmqwm šb'th). Just as [in the case of a substitute], those that are unblemished may not be redeemed, while those that are blemished are redeemed, so here [in the case of what is purchased as second tithe], those that are clean may not be redeemed, [while] those that are unclean may be redeemed."

T. 2:17 (p. 256, ll. 101-106)
(E: M. Tem. 2:3)

T. cites M. 3:10A-C and then gives its own version of the debate. The authorities of A now criticize Judah for being more strict in regard to what is purchased to replace produce separated as second tithe than he is for the originally designated produce itself. While produce originally separated as second tithe may be deconsecrated if it becomes unclean (M. 3:9), Judah does not allow this same leniency in the case of items purchased as second tithe. He responds to A's criticism by pointing out that there are other cases in which what is consecrated in place of another item is in fact treated more stringently than the item it replaces. If, for example, one dedicates an animal for use on the altar and the animal has a permanent blemish, the animal is not deemed to be consecrated. Yet if an animal is made a substitute (temurah) for a dedicated animal, that substitute is in its consecrated status even if it has a permanent blemish. The substitute in this case is treated with greater stringency than the animal it replaces. The same holds true, Judah argues, for produce which is purchased to replace what the farmer separated as second tithe. F. responds by turning Judah's own example against him. An animal designated as a substitute, and which is blemished, may be deconsecrated. It follows that produce purchased as second tithe and which becomes unclean should also be capable of being deconsecrated.

3:11

A. A deer which one purchased with money [in the status of] second tithe and which died,
B. is to be buried with its hide.[23]
C. R. Simeon says, "It is to be redeemed."
D. [If] one purchased it alive and slaughtered it, and it [subsequently] became unclean,
E. it is to be redeemed.
F. R. Yosé says, "It is to be buried."
G. [If] one bought it [when it was already] slaughtered, and it [subsequently] became unclean —
H. behold, it is [treated] like produce [in the status of second tithe which becomes unclean and is redeemed].

M. 3:11
(B-C: cf. M. Tem. 7:3)

Produce in the status of second tithe which becomes unclean is sold, and other foodstuff is purchased in its stead (M. 3:10 B). The question now is whether or not animals are analogous to produce such that if the farmer purchases an animal as second tithe and it becomes unclean, it too is sold and replaced with other edibles. In dealing with this problem, the pericope distinguishes between living animals and animals that have been slaughtered and are thus ready to be used as food. According to G-H, the analogy does hold if the farmer purchases a slaughtered animal, that is, meat. Like produce, meat is a food item and so, the authority behind G-H reasons, is to be treated in the same way. It is sold if rendered unclean. In A-C and D-F, however, the farmer purchases a living animal for use as second tithe. While the farmer intends to slaughter the animal, it is not a food at the time it is rendered unclean. Under dispute is whether or not the farmer's intention to use the animal as food nonetheless imposes on it the status of food, so that if it becomes unfit for use it is sold. This point is brought out by the disputes in A-C and D-F.

In A-C the animal which the farmer purchases dies before it is made fit for use as food. The dead animal as nebelah[24] is forbidden for consumption. Since the animal was never in the status of food, B rules that it is not to be treated like produce. That is, the farmer may not sell the unclean carcass and replace it with other edibles. Simeon disagrees. The farmer bought the animal with the intention of using it as food. It follows, Simeon holds, it must be treated as such.

D-F presents a case in which the farmer purchases a living animal and then properly slaughters it, thereby making it ready for use as food. This meat subsequently becomes unclean, giving us a case like that of G-H. The anonymous rule declares that the meat, which is in the status of food, is sold and replaced with edibles. Yosé (F) disagrees, claiming that the unclean venison is buried. His point, presumably, is that meat is never analogous to produce. Laws which apply to fruit, therefore, cannot be made to apply to animals. If meat which the farmer prepares as second tithe becomes unclean, therefore, it may not be sold. Yosé, then, clearly stands in opposition to the presupposition which generates the rest of the pericope.

3:12-13

A. He who lends out jugs [to hold wine[25] to be sold as] second tithe,
B. even if he corked them [i.e., the jugs]
C. [he] does not acquire [the status of] second tithe for the jugs.
D. If he poured [wine] into the jugs without specifying (stm) [that the jugs were not being sold but only lent][26] —
E. before he corked them [he] does not acquire [the status of second] tithe [for the jugs].
F. After he corked them, [he] does acquire [the status of second] tithe [for the jugs].
G. Before he corked them, [consecrated liquids contained in

them] are neutralized [in a mixture of] one hundred and one [parts of unconsecrated liquid].

H. After he corked them, the jugs render consecrated any number [of corked jars containing unconsecrated wine with which they become mixed].

I. Before he corked them, he may remove heave-offering from one jug for all [the jugs in the heap].

J. After he corked them, he removes heave-offering from each jug [individually] (mkl 'ḥt w'ḥt).

M. 3:12 (G: cf. M. Ter. 4:7; H: cf. M. Or. 3:7)

K. The House of Shammai say, "[If a wine merchant does not want a corked jug to be purchased along with the wine it contains], he opens [the jug] and pours [the wine] into the vat."

L. And the House of Hillel say, "[The merchant] opens the jug, but he does not need to pour [the wine back into the vat]."

M. To which [case] does this apply?

N. In a place where [jugs] are normally sold sealed.

O. But in a place where they are normally sold open, the jug does not [remain] unconsecrated [i.e., it is purchased along with the wine] [= M. 1:4M-N].

P. But if he [i.e., the merchant] wanted to impose a stringency upon himself and sell [the wine] only by [exact] measure —

Q. the jug [remains] unconsecrated.

R. R. Simeon says, "Also: he who says to his friend, 'I sell to you this cask [of wine] except for its container (reading with most MSS mknknh)' —

S. "the jug [remains] unconsecrated."

M. 3:13

A farmer is purchasing wine for use as second tithe. Along with the wine he receives the jug in which it is contained. Is this jug also deemed to be in the status of second tithe? If it is, then it must be resold and food purchased in its stead. The status of the jug, these pericopae declare, depends on the merchant's intention in giving it to the buyer. If the grocer intends only to lend the jug out, it remain unconsecrated. If, on the other hand, the seller deems the jug to be part of the sale, it does become consecrated. This notion is articulated in the two rules at A-C and D-F. K-S is a long secondary expansion of D-F. The rules at G-H and I-J are formally parallel to D-F, giving us a triplet.

In A-C the merchant makes it known that he does not intend to sell his wine jugs. Rather, he loans them out for the convenience of his customers. Since the jug is not part of the sale of the wine, it does not take on the status of second tithe from the coins used to purchase the wine. D-F presents a more difficult case in that the merchant has not made his intention known. We must turn, therefore, to other signals of his intention. These we find in his decision to cork the jugs or to leave them uncorked. If he has left the jugs open, the wine in them will have to be used in the near future. The farmer who buys it will soon be able to return the empty jug and the jug is therefore deemed to be lent out. If the merchant sells the jug, however, we assume that he expects that the buyer will not return it. The farmer who purchases the product has purchased the jug as well and must resell it.[27]

G-H and I-J introduce other cases in which the fact that a jug is sealed or unsealed has legal ramifications. In G-H a jug containing consecrated wine becomes mixed in with a batch of other jugs containing unconsecrated wine. If all of the jugs are open, we consider the wine contained in them to intermingle, forming one large batch. Consequently, if consecrated wine is in such a batch at the ratio of one part consecrated wine to one-hundred parts secular wine, the consecrated portion loses its special status as it would in any other mixture.[28] If, on the other hand, all the jugs are sealed, no intermingling takes place. The wine in each jug retains its discrete identity. Since we do not know which jug contains the consecrated wine, the entire collection is treated as though it were in a consecrated status.[29] In I-J we have a similar situation. Untithed wine is poured into a number of different jugs. Before the jugs are sealed, the wine in them is still considered to be part of the larger original batch. Wine from one jug may therefore be designated as heave-offering or tithe for the rest. Once the jugs are sealed, however, each small quantity is considered to be a separate batch and tithes must now be removed from each jug individually.[30]

The Houses' dispute in M. 3:13K-L carries forward the ruling at M. 3:12D-F. There we saw that if the merchant has not specified that he is lending the jugs and, after filling them with wine, seals them, the jugs are deemed to be purchased along with their contents. What must the merchant do if he subsequently decides not to sell the jugs after all? The Shammaites hold that he must undo the entire bottling process. That is, he must pour the wine back into the vat and, after announcing his intention to lend the jugs, refill them again. The Hillelites declare that he need do no more than simply uncork the jugs. As we have seen in M. 3:12D-E, uncorked jugs are deemed not to be sold along with the wine they contain. The Hillelites, then, are concerned only with the physical status of the jugs. The Shammaites, in contrast, are interested in the merchant's intentions throughout the bottling process.

M-O now qualify the law developed by the Houses. We are told that the preceeding rules are applicable only in a place in which jugs of wine are normally sold sealed. In such a place, purchasing open jugs is taken as a sign that the purchaser does not intend to keep the jugs. In a place in which jugs of wine are normally sold open, however, the fact that the farmer buys uncorked jugs has no special significance. The containers are therefore deemed to be sold.[31] At this point, M. introduces there two exceptions to the above rule that in a place in which wine is normally sold in open jugs, open jugs are deemed part of the purchase. The merchant in F-G sells wine in open containers so that he can easily measure the exact amount of produce he is selling. Since it is clear that the wine is in such jugs for the convenience of the merchant, they are not deemed to become part of the sale. In R-S the seller clearly states that the container is not included in the transaction. This final ruling brings us back to the case of M. 3:12A-C, in which the seller's intentions are clearly unknown from the outset. In this way the redactor signals the end of his essay.

A. <u>He who lends out jugs [to hold wine purchased with coins in the status of] second tithe,</u>

B. <u>even if he corked them —</u>

C. <u>[he] does not acquire [the status of] second tithe [for the</u>

jugs] [M. 3:12A-C].

D. To which [case] does this apply?

E. [To jugs] of wine.

G. But [in regard to jugs] of (1) brine, and of (2) vinegar, and of (3) fish-brine, and of (4) oil, and of (5) honey,

G. (1) whether he corked them or did not cork them [the jugs] have acquired [the status of] second tithe [M. 3:12F];

H. (2) whether he corked them or did not cork them [consecrated liquids in the status of heave-offering which are in them] are neutralized [in a mixture of] one [part consecrated liquid] to one-hundred [parts unconsecrated produce] [cf. M. 3:12G];

I. (3) whether he corked them or did not cork them, [liquids in the status of heave-offering which are in them] render consecrated any number of jars containing [unconsecrated produce in jugs with which they become mixed] [cf. M. 3:12H].

J. Under what conditions ('ymty) did they say [the jugs] acquired [the status of] second tithe [cf. M. 3:12F]?

K. When all [its contents are in the status of second] tithe.

L. But if he left in it (hpqyd) a quarter-log of unconsecrated [liquid],

M. whether he corked it or did not cork it,

N. the jug has not acquired [the status of second] tithe.

O. Said R. Simeon b. Eleazar, "The House of Shammai and the House of Hillel did not disagree about one who presses (dwrk) [grapes] in a cask [that to indicate his desire to sell only the wine and not the jug] he [needs only to] open [the jug] and does not need to pour [the wine out].

P. "About what did they disagree?

Q. "About one who presses [grapes] in a vat.

R. "For the House of Shammai say, 'He opens [the jug] and he pours [the wine] into the vat.'

S. "And the House of Hillel say, 'He opens [the jug] and he does not need to pour [the wine back into the vat] [M. 3:13A-B].'"

T. 2:18 (p. 256, ll. 106-113)

T. cites and qualifies the rules of M. 3:12-13. T.'s glosses of M. 3:12 appear in two blocks of material: D-I and J-N. O-Q deals with the Houses' dispute of M. 3:13A-B.

According to M., cited in A-C, jugs acquired from a merchant who normally lends out containers do not take on the status of second tithe. D-F now limit the application of this rule to jugs containing wine. The items listed in F are either too hard to remove (e.g., honey) or impart a strong odor to the pottery (e.g., fish-brine) such that the jug cannot be reused. We assume in these cases that the merchant does not expect the jug to be returned and so considers it sold to the buyer. Whether the jug was corked or left open is now irrelevant. The implications of this qualification are spelled out in G-I. In each case, T. has simply borrowed an apodosis from M. 3:12F-H and attached its own protasis: "whether he corked them or did not cork them . . . . " As it now stands, H and I are mutually exclusive. H tells us that the consecrated produce is neutralized in the larger batch while I declares that the entire collection becomes consecrated. GRA, following y. M.S. 3:10, deletes the present apodosis in I and substitutes for it the apodosis of M. 3:12I: "he may remove heave-offering from one jug for all [of them]."[32] With GRA's emendation, H and I are perfectly consistent with each other. In each case the collection is treated like a single batch. That is, consecrated items which are mixed in with it are neutralized (H) and tithes are removed from any part of the wine for the entire batch (I).

J-N refer to a case in which the jug is deemed to be purchased as second tithe along with its contents (M. 3:12F; F-G above). T. now rules that this is so only if all of the liquid in the jug is purchased for use as second tithe. If the purchaser specifies that only some of the wine is to be purchased with his consecrated coins, then only that wine acquires the status of second tithe. It follows that what has not been mentioned, the rest of the wine as well as the container, do not become consecrated.

The final unit of T., O-S, comments on the Houses' dispute by giving a case in which the Houses will no longer disagree. As we recall, the farmer has poured wine into jugs and sealed them without specifying whether the jugs are to be sold or lent. The law assumes that in this case the jugs are sold. If the merchant then decides only to lend the jugs to the buyer, the Hillelites claim that he simply uncorks them, while the Shammaites rule that he must start the bottling process over from the beginning by returning the wine to the vat. Simeon b. Eleazar claims that the Houses will be in agreement if the wine was originally pressed and sealed in the same cask. The wine is already in the original vat and so all that he is required to do is uncork it.

CHAPTER FOUR
MAASER SHENI CHAPTER FOUR

The bulk of the chapter, M. 4:1-8, regulates the price a farmer may charge for consecrated produce. It is this money which he takes to Jerusalem and uses there to purchase food he will eat as second tithe. It is essential, therefore, that this money represent the value of the originally consecrated produce it replaces. A concluding unit, M. 4:9-12, takes up a second, unrelated issue. Its concern is with how one treats produce and coins the status of which is in doubt. Characteristically, M.'s redactor has placed a discussion of doubts at the close of a major thematic unit, in this case the tractate dealing with second tithe. Let us now briefly outline the main points of each of these parts of the chapter.

M. 4:1-8's interest in establishing the selling price of second tithe produce assumes that produce has no intrinsic material value. It is worth only what consumers are willing to pay for it. It follows that the value of consecrated produce is determined by the price that its same sort of produce commands on the market. The laws of the chapter, accordingly, describe how the farmer determines the market value of his second tithe produce. The discussion proceeds in three steps. First we determine how the selling price is established when the produce is sold on the open market (M. 4:1-2). Concern then shifts to cases in which the farmer exchanges the produce for his own coins (M. 4:3-5). Finally, (M. 4:6-8), it examines cases of uncertainty in which the market value of the produce shifts before the sale is completed.

M. 4:1 states the basic proposition of the chapter: that consecrated produce is sold for the same price as is unconsecrated produce of its same type. This proposition is then the subject of two further developments in M. 4:2. The first of these is that the farmer may calculate the price such that he is able to sell his consecrated produce competitively. This he does by selling the produce at wholesale prices while accepting coins in payment at their premium value. This insures that the farmer will find sufficient buyers for his goods. Secondly, M. establishes a mechanism for setting a price for the produce if similar produce is not sold on the market. In such a case, the farmer auctions his consecrated food for the highest of at least three bidders (M. 4:2F-G).

The second unit of this discussion, M. 4:3-5, turns to a related topic, the case in which a farmer transfers the status of second tithe from the produce to his own coins. M. follows Lev. 27:31 in stating that in such a case the farmer must pay an added fifth over the normal selling price of the food. The discussion here concerns how this scriptural law is applied. M. 4:3A-D declares that this fee is treated strictly as a surcharge and not part of the basic selling price of the produce. Thus the added fifth is not taken into account when the farmer is entering a bid to establish a price for his produce as in M. 4:2. M. 4:3-5 develop the notion that this added fifth is paid only when the farmer transfers the status of second tithe from his own produce to his own coins. He thus can avoid payment by arranging to have a member of his household buy the produce. Since another person is buying the produce, the farmer is not liable for the added fifth even though both the coins and the produce remain in his domain.

We turn in M. 4:6-8 to the third topic of concern, the special problems created by constantly changing prices. Specifically, M. 4:6 describes a situation in which the market value of produce in the status of second tithe changes after the purchaser agrees to a selling price but before he makes payment. The result is that the amount of money which the purchaser will pay to the farmer no longer reflects the market value of the consecrated produce he is buying. If the produce has decreased in value, the purchaser ends up giving the farmer more money than is necessary to deconsecrate the produce. In this case, M. declares that the extra money remains unconsecrated. Conversely, if the produce has increased in value, the money used to purchase it will be insufficient fully to deconsecrate it. M. rules that in this case some of the produce retains its consecrated status. The purchaser must be sure to take this portion of the produce to Jerusalem or to deconsecrate it. This discussion leads into a secondary issue, debated by Yosé and Judah in M. 4:7. The question is whether the payment of coins is sufficient to deconsecrate produce or whether the purchaser must also make an oral declaration declaring his intention. Yosé declares that such a declaration is unnecessary while Judah holds that it is required.

M. 4:8 concludes the discussion of shifting market prices, begun at M. 4:6, by considering the value at which the consecrated coins themselves are spent in Jerusalem. M. 4:8A-F state that the purchasing power of these coins is not fixed but changes according to local market conditions, in this case Jerusalem. In this way, M. 4:8 restates the same principle introduced at M. 4:1-2. M. 4:8G-I make the related point that since these coins have no set market value, they may be deemed fully deconsecrated even if a small portion of their value remains unspent.

The second part of the chapter, M. 4:9-12, introduces considerations of doubt and with this concludes M.'s discussion of second tithe. Under consideration are cases in which coins or produce are found but it is not known whether or not they are in the status of second tithe. According to Judah (M. 4:10) and the anonymous opinions of 4:11 and 4:12D-I, if there is any evidence that they are in a consecrated status, they must be treated as such. Yosé, on the other hand, claims that in all cases of doubt, the coins or produce are deemed to be unconsecrated.

4:1

A. One who carries (hmwlyk) produce [in the status of] second tithe from a place [where it is] expensive to a place [where it is] cheap,

B. or from a place [where it is] cheap to a place [where it is] expensive,

C. redeems it according to the market price (š^{c}r) of his [current] location.

D. One who brings produce [in the status of second tithe] from the threshing floor to the city,

E. or (w) jugs of wine from the vat to the city —

F. the increase in value [accrues to the] second [tithe]

G. and the expenses [involved in transporting the produce] come out of the farmer's pocket (mbytw).

M. 4:1

The farmer is selling his consecrated produce in accordance with Deut. 14:23. The money he receives from the sale is used in Jerusalem to buy food to be eaten in place of the original. The problem is to insure that the full value of the produce is transferred to the money so that the money the farmer brings to Jerusalem is equal in value to what he originally designated as second tithe. M. declares that this value is determined by the market-price of the produce at the time and place of sale. The reason is that, for M.'s authorities, produce is worth only what consumers are willing to pay for it as food. It has no intrinsic material value. It follows that its value at the time of redemption is the price that its same type of food commands on the local market. This is the point of A-C. D-G illustrate and augment this point. The farmer brings his second tithe to the city market, where it fetches a higher price than it would in the country. The produce is sold at this higher price, as A-C have told us. G makes the obvious point that the farmer may not use part of the money he receives to cover the expense of bringing the produce to the higher-priced market. Rather, its full amount must be brought to Jerusalem and spent there on food to be eaten as second tithe.

A. One who carries produce [in the status of] second tithe from a place [where it is] expensive to a place [where it is] cheap,

B. or from a place [where it is] cheap to a place [where it is expensive],

C. redeems it according to the market-price of his [current] location [M. 4:1A-C].

D. Said R. Joshua b. Qorḥa, "To which [case] does this apply? [To a case in which it has been separated] from doubtfully tithed produce. But if it was separated from certainly untithed produce, it is redeemed at the more expensive market-price (kš^{c}r hywqr).

T. 3:1a (p. 257, ll. 1-3)
(Cf. y. M.S. 4:1)

T. cites and glosses M. 4:1A-C. According to M., the farmer may bring his produce in the status of second tithe to a low-priced market, even though he receives less money in exchange for the consecrated produce. Joshua, D, now takes away what M. has given. According to him, this rule applies only to produce removed as second tithe from produce which may already have had tithes removed.[1] What is now designated as second tithe may not be in that status at all. This type of produce may be sold cheaply. Produce certainly in the status of second tithe, however, must be deconsecrated at its highest market value. His ruling certainly will benefit the merchants of Jerusalem.

A. <u>One who brings produce from the threshing-floor to the city</u> [M. 4:1D],

B. [and then] separated heave-offering and gave it to a priest, first tithe and gave it to a Levite, poor man's tithe and gave it to a pauper,

C. does not deduct (<u>mḥšb ᶜmhn</u>) the cost of transporting [the produce to the city, thereby reducing the amount of these gifts].

D. But (<u>'bl</u>) if he designated these [gifts] at the threshing-floor, lo, he does deduct the cost of bringing [the produce to the city].

E. One who sells to his friend produce and says to him, "The produce I sold you is untithed,"

F. [if he sold him] meat [and says to him], "It is meat from a firstling";

G. [if he sold him] wine [and says to him]: "It is libation wine" —

H. that which they have eaten is eaten

I. [but] let him [the seller] return their money.

J. R. Simeon b. Eleazar says, "[As regards a thing] which is disgusting to eat (<u>hnpš qṣh bw</u>) [such as <u>terefah</u>, <u>nebelah</u>, or creeping things],[2] that which is eaten [by mistake] is eaten, but [the seller] returns his money.[3]

K. "And [as regards] a thing [which is not proper to eat but] which is not disgusting to eat [such as untithed produce, firstlings, and wine used for libations], [the seller] deducts from the money [he must return the value of the produce which was eaten]."

T. 3:12 (p. 260, ll. 37-44)
(E-I: b. Bek. 37a)

In A-D we assume that it is to the advantage of the recipients of heave-offering and tithes to have their produce brought to the city. The question before us is who must pay the cost of transporting it. As we shall see, the answer depends on who owns the produce. In A-C the produce has not yet been designated as heave-offering and tithe and therefore, according to

Rabbi, still belongs to the farmer. He, therefore, must bear the cost of bringing it to the city. In D, on the other hand, the produce is already designated as heave-offering or tithe when it is being conveyed to the city. Since it now belongs to the priests, Levites or paupers, they must bear the cost of transporting the produce to the city.

In E-I proscribed produce is mistakenly sold to one who may not consume it. E-G hold that the transaction is undone, with the seller refunding the purchase price and the buyer returning whatever he has not already consumed.[4] Simeon b. Eleazar holds that the amount of money to be refunded depends on the nature of the produce. If the produce is by nature disgusting, the entire purchase price is refunded. We assume the consumer received no benefit from it at all. If, however, the produce was in fact usable, the merchant needs to give a refund only for what is actually returned by the consumer. The buyer did in fact benefit from the produce he ate.

4:2

A. They redeem [produce in the status of] second tithe according to its lowest selling price (<u>kš^cr hzl</u>):

B. [the rate] at which the shopkeeper buys and not [the rate] at which he sells:

C. [the rate] at which the moneychanger sells [small change] (<u>pwrt</u>) and not [the rate] at which he buys [small change] (<u>msrp</u>).

D. And they do not redeem [produce in the status of] second tithe by estimating [its worth] (<u>'kśrh</u>).

E. [Produce] the price of which is known is redeemed according to the [valuation of] one [buyer: so T. 3:5] (<u>ᶜd</u> lacking in most mss.),

F. and [an item] the price of which is not known is redeemed according to [the valuation of] three [buyers].

G. For example: wine which has formed a film (<u>šqrm</u>, most mss. read <u>šqśś</u>: which turned sour) or produce which has begun to rot or coins which are rusty.

M. 4:2 (A-C: y. M.S. 4:2;
F: M. San. 1:3; E-G: b.
San. 14b)

Although the farmer must sell consecrated produce for its full market price (M. 4:1), M.'s authorities want to make it possible for the farmer to sell the produce at a competitive price. The assumption is that the farmer is selling the produce because he is unable to bring it all to Jerusalem. It therefore is important that he be able to sell it all. The rules before us explain how the farmer can set a competitive price for his consecrated food while technically receiving in exchange a price in accord with local market conditions. The rules are of two sorts. In the first place, the farmer may sell the produce at its wholesale price (B). In this way he can offer it for a price that is lower than that found at neighboring stalls. In addition, he may accept the purchaser's money at an inflated value. This he does by according to the purchaser's coins the higher value they would have when sold by a moneychanger (C).[5]

D introduces a secondary issue, spelled out at E-G. The point is that consecrated produce must be sold at a price set by the market and not by a price arrived at privately by the farmer and a prospective buyer. This is no problem when the type of produce being sold has a known market price. Since the price is public knowledge, the farmer can take the buyer's word for what the price should be (E). A difficulty arises, however, if no similar produce is available in that market. Now there is no market price on which to base a selling price. F declares that, in this case, a market price is created by auctioning off the produce. A minimum of three bids is assumed to constitute a market for this purpose and the consecrated produce may be sold for the highest offer. G supplements F by supplying examples of items which, being damaged, have no preestablished market value, and so must be auctioned.

E. They redeem produce [in the status of] second tithe according to its lowest selling price:

F. [the rate] at which the shopkeeper buys and not [the rate] at which he sells [M. 4:2A-B].

G. R. Simeon b. Eleazar says, "[At the rate] at which the shopkeeper buys [produce] of that same type."

T. 3:1b (p. 257, ll. 3-5)

T. cites M. 4:2A-B, supplying a gloss at G. Simeon specifies that the value of produce to be deconsecrated is set according to the market price of the same kind of produce. This prevents the farmer from selling his consecrated produce at the price accorded to cheaper items.

A. [A farmer] who sells (pwrṭ) a silver dinar [in the status of] second tithe [for small change] —

B. [does so at the higher value it has] when the shopkeeper buys [produce wholesale] and not [at the lower value it has] when he [i.e., the shopkeeper] sells [produce retail] [M. 4:2B].

T. 3:2 (p. 257, ll. 5-6)

A. One who buys a gold dinar [in the status of] second tithe [with small change] —

B. [does so at the high price] at which the moneychanger sells [gold dinars] and not [at the low price] at which he [i.e., [the moneychanger] buys gold dinars [M. 4:2C].

T. 3:3 (p. 257, ll. 6-7)

A. "One would take [a gold dinar] from one place to another and earn a profit (ytr ^{c}l šywyw) of a roba^{c} [one half dinar]," the words of R. Judah.

B. R. Eliezer says, "Up to [a silver] dinar."

T. 3:4a (p. 257, ll. 7-9)
(Cf. y. M.S. 4:1)

T. carries forward the principle of T. 3:1a, now applying it to coins. It requires the farmer to sell coins in the status of second tithe at their highest market price. The result is an increase in the amount of money which the farmer receives and eventually takes to Jerusalem. In making this claim, T. reverses the position taken up by M. 4:2, which declares that the farmer may sell his consecrated merchandise cheaply so that it will be easily sold. We have a doublet at T. 3:2-3 with autonomous material added at T. 3:4a.

We deal first, T. 3:2, with a case in which the farmer is selling a coin of large denomination. The coin is given the value it would have in the hands of a shopkeeper buying produce wholesale. At wholesale prices the dinar purchases more produce and therefore has increased value. By selling the dinar at this higher value, the farmer increases the value of coins he receives in return. The same result is achieved in T. 3:3. A householder is buying from a farmer a gold dinar in the status of second tithe, giving the farmer small change in return. The householder pays for the gold dinar the price a moneychanger would charge when selling it. Since a moneychanger demands a high price for coins he is selling, the householder must pay a premium price for the gold dinar. The result again is an increase in the total value of coins being consecrated.

T. 3:4a serves as an appendix to the foregoing, indicating that the relative value of coins fluctuates from market to market. It is therefore possible for the farmer to receive more money for his consecrated coins simply by selling them in a different location. From Eliezer's comment that the value of the coin can increase by a dinar, it is probable that the coin in question is not itself a silver dinar but a gold dinar, which is equivalent to twenty-five silver dinars.[6]

C. [As regards produce in the status of] second tithe [which] has no value —

D. it is required (dyn hw') that he says, "It and its added fifth[7] are deconsecrated with this issar (one twenty-fourth of a dinar)."

T. 3:4b (p. 257, ll. 9-10)
(Cf. y. M.S. 4:3)

The farmer has produce in the status of second tithe which has no market value. Since some sanctity does inhere in the produce, it is deconsecrated for some small amount of money. I do not know why T. specifies an issar.

A. [Produce in the status of] second tithe the value of which is not known is redeemed according to [the valuation of] three bidders (lqwḥwt) [cf. M. 4:2F],

B. and not according to [the valuation of] three who are not bidders,

C. even if one [of the bidders] is a non-Israelite,

D. even if one [of the bidders] is the owner [of the produce] (reading singular for b^{c}lym).

E. They force the owner to open [the bidding] (lptwḥ r'šwn).

F. If he says, "Lo, it is mine for an issar," he may retract [his bid if it is shown to be unreasonable].

G. This [is a case in which the law] is more stringent in regards to dedicated [items] than [in regard to produce in the status of] second tithe.

T. 3:5 (p. 257, l. 10 — 258, l. 13) (A: T. San. 1:2, T. Arak. 4:2; A-D: y. M.S. 4:2)

A restates M. 4:2, which declares that three bidders can establish the price of produce not otherwise found on the market. B-D gloss A. It rules that any prospective buyer may enter a bid, even the owner himself. If the owner is one of the bidders (E-G), he opens the bidding. If his bid turns out to be too low, he is allowed to retract it, accepting for his produce the price set by the others.

It is not clear to what stringency G refers. Lieberman suggests that it has to do with the owner's right to retract his bid.[8] He finds a parallel to our law in M. Arak. 8:1. There, a farmer dedicates a field to the Temple at a time in which the Jubilee is not in effect. The farmer can regain ownership of the field only by redeeming it. The value of the field is established by three bidders, and if the owner is one of the three bidders, he enters his bid first. No mention is made of his right to retract his first bid. The law of dedication therefore appears to be more stringent than that for second tithe because, in the case of a dedicated field, the owner is not given a second chance to redeem his property.

A. They do not redeem [items in the status of] second tithe by estimating [their worth] [M. 4:2D].

B. How so?

C. [If] one had rotting produce or rusty coins, [cf. M. 4:2G]

D. he should not say, "How much is one willing to pay for this collection [of coins], and how much is one willing to pay for this heap [of produce]?"

E. Rather he says, "How much [of this] produce is one willing to purchase for a sela^c, and how many [of these] coins is one willing to purchase for a [silver] dinar?"

F. "How much [weight] may [a coin] lack until [produce in the status of] second tithe may not be deconsecrated with it?

G. "[In the case of] a *sela*^{c}, four *issars* [one twenty-fourth of a *sela*^{c}]; [in the case of] a *dinar*, one *issar* [one twenty-fourth of a *dinar*]," the words of R. Meir.

H. R. Judah says, "[In the case of] a *sela*^{c}, four *dupundia* [one twelfth of a *sela*^{c}]; and [in the case of] a *dinar*, one *dupundium* [one twelfth of a *dinar*]."

I. R. Simeon says, "[In the case of] a *sela*^{c}, eight *dupundia* [one sixth]; and [in the case of] a *dinar*, two *dupundia* [one sixth]."

J. [If the coin lacks in weight] more than this, it [is used for] deconsecrating according to its actual [weight] [and not its face value].

K. [This holds true] for a *sela*^{c} until [it weighs only] a *sheqel* [one half *sela*^{c}] and for a *dinar* until [it weighs only] a *rova*^{c} [one half *dinar*].

L. [If the coin weighs] even an *issar* less than this, one is not authorized to deconsecrate [produce] with it.

M. If a blank [a coin the features of which have rubbed away] circulates as a *sela*^{c}, but as a [recognizable] coin [it would circulate as] a *sheqel*, or a blank circulates as a *sheqel*, but as a coin [it would circulate as] a *sela*^{c} —

N. [its value as regards deconsecrating produce in the status of] second tithe is the value [it would have as a recognizable] coin.

T. 3:6 (p. 258, ll. 13-22)
(F-I: cf. M. B.M 4:5;
F-J: cf. T. B.M. 3:17;
K-N: y. B.M. 4:4)

T. supplies rules for determining the price of produce or coins which are damaged. We deal first, in B-E, with items which have only little market value. Rusty coins are not accepted as legal tender and rotten produce cannot be eaten. The farmer sets a minimum price which he will accept for these items. He then lets the consumer pick an amount of produce or coins he is willing to purchase at that price. F-N deals with coins which have become worn. These are different from the rusty coins discussed in B-E since a value can be set for these coins on the basis of the amount of metal still contained in them. Their market value is determined as follows. If the coin has lost only a small proportion of its metal (G-I), it is exchanged at face value. If a significant amount of metal has been lost, but less than half, the coin's value is set in accordance with the proportion of metal that remains (J-K). Once more than half the coin's metal has been worn away, it is deemed to be worthless. The farmer may not use such coins to deconsecrate produce in the status of second tithe.

M-N turns our attention to coins the metal content of which is largely intact but the features of which have been rubbed smooth. In this case the denomination of the coins is not evident.[9] If it was originally a sheqel but now appears to be a sela^{c}, or vice versa, it is accorded the full value it had when minted.[10] This is so because the coin still contains the same amount of metal as when it was first issued.

A. One who sells produce to his fellow and it turns out that the produce is untithed
B. runs after him to make him properly remove tithes [from the produce] (ltqnn).
C. [If] he does not find him —
D. if it is known that [the untithed produce] [still] exists [i.e., it has not been consumed] [the seller] must separate tithes for it [from his own produce];
E. if it is known that [what he sold] does not exist [any longer], he does not need to separate tithes for it;
F. if there is doubt as to whether or not [what he sold] still exists (spq šhn qyymyn spq š'ynn qyymyn), he separates tithes for it and he [also] designates tithes [for what he separated, in case the separated produce is not in the status of tithe at all but remains in the status of untithed produce].

T. 3:7 (pp. 258, l. 22-259, l. 25) (y. Dem. 7:1)

G. One who sells produce to his friend, and the seller says, "I sold you untithed produce," and the buyer says, "You sold me properly tithed [produce] (mtwqnyn)" —
H. they force the seller to properly remove tithes [for the buyer's produce] from his own [produce].
I. Lo, if the royal authorities (mbyt hmlk) took [produce directly] from the threshing floor —
J. if [the produce was taken] on account of a debt [the farmer owed], he must separate tithes for it.

K. If [the produce was] confiscated [against his will], he need not separate tithes for it.

T. 3:8 (p. 259, ll. 25-28)
(I-K: cf. b. Git. 44a: b. Hul. 131a)

Untithed produce may not be eaten until heave-offering and tithes have been separated for it. A consumer now buys untithed produce but for some reason is not expected to separate the priestly gifts. The responsibility to do so, as we shall see, devolves on the farmer or merchant who sold the untithed produce. Formally, we have a doublet at A-F and G-H followed by an autonomous law at I-K.

We begin with a case in which the merchant unknowingly sells untithed produce and later realizes his mistake. He runs after the purchaser in order to tell him to separate heave-offering and tithes, but is unable to find him. If the merchant has reason to believe that the produce has not yet been consumed, he himself designates heave-offering and tithes for it. If, however, he is sure that the untithed produce already is eaten, he does not need to designate these gifts. The produce for which tithes were once owed no longer exists. If the merchant is unsure as to whether or not the produce still exists, he separates heave-offering and tithes for it anyway. What he now separates, however, may not enter the status of heave-offering or tithes at all, since the produce for which he designates it may no longer exist. He therefore must designate heave-offering and tithes as well for what he has just separated.

In G-H the merchant has located the person who purchased the untithed produce. The purchaser, however, refuses to separate the priestly gifts, claiming that what he bought was properly tithed. Since the purchaser refuses to separate tithes, the responsibility to do so falls on the merchant, as in A-F.

In I-K we see that only if the farmer receives benefit from his crop is he obligated to separate heave-offering and tithes for it. Governmental authorities confiscate produce before the farmer has had a chance to remove priestly gifts. If the produce is taken to repay a debt, the farmer receives some benefit from his produce. He must separate the priestly gifts for it as he would if he sold the produce (A-G). If the produce is simply taken away, with no benefit accruing to the farmer, he is under no obligation to remove the priestly gifts (K). This is so because he was never able to use the produce for his own purposes.

A. "One who eats produce [in the status of] second tithe [as though it were unconsecrated, i.e., outside of Jerusalem or in a state of uncleanness],

B. "whether this was by mistake or on purpose,

C. "appeals to heaven [for forgiveness] (yz^c^q lšmym) [i.e., he can not make up for the loss]," the words of Rabbi.

D. Rabban Simeon b. Gamaliel says, "[If he did so] by mistake, he appeals to heaven; [if he did so] on purpose, its value is repaid (yḥzr dmym lmqwmn)."

T. 3:9 (p. 259, ll. 28-30)
(y. M.S. 1:1)

E. "One who spends ('wkl) coins [in the status of] second tithe, [on produce which is then eaten as though it were unconsecrated]

F. "[if he does so] by mistake, appeals to heaven [for forgiveness];

G. "[if he does so] on purpose, its value is replaced [with other produce to be eaten as second tithe] [cf. M. 1:5-6]," the words of Rabban Simeon b. Gamaliel.

H. Rabbi says, "Whether [he does so] by mistake or on purpose, its value is repaid.

T. 3:10 (p. 259, ll. 30-32)
(y. M.S. 1:1)

We have two alternate positions as to whether or not a farmer who misuses produce or coins in the status of second tithe must replace their value. Rabbi holds that the farmer does repay the value of misused coins but not the value of misused produce. Simeon b. Gamaliel, on the other hand, declares that if the farmer purposely misappropriated either of these items, he must replace their value. These two views have been woven together to form a perfectly balanced doublet.

Rabbi's position can be understood in light of the principle adduced at T. 3:7. There we saw that a farmer may designate produce as heave-offering or tithes to replace other produce, as long as the original produce still exists. If the original produce is eaten, the farmer may not consecrate other produce in its stead. The same restriction is now applied here. If the farmer improperly eats produce designated second tithe, he cannot replace it since it no longer exists. He has no choice then but to pray for forgiveness for having misused consecrated produce. The case, however, is different for coins. The coins, although misspent, continue to exist. It is therefore possible for the farmer to designate other coins to take on the status of second tithe in their place, and he must do so.

Simeon b. Gamaliel, as we noted, bases his ruling on the intention of the farmer. If the farmer was aware that he was misusing consecrated items, he must replace their value. This prevents him from gaining any benefit from purposely violating the laws of second tithe. If his misuse was an honest mistake, however, Simeon b. Gamaliel rules that he does not need to repay the value of the misappropriated items. The farmer, however, must pray to heaven for forgiveness since he did in fact misuse consecrated goods.

A. "One who eats untithed produce [belonging] to his friend, pays back the value of [the entire batch including the value of what should have been removed as heave-offering and] tithes from the untithed produce," the words of Rabbi.

B. R. Yosé, the son of R. Judah, says, "[He pays only] the price of the unconsecrated produce in it."

C. An Israelite who ate his produce while it was in its untithed status (bṭblw) —

D. even though he is answerable in [the eyes of] heaven, the priest has no claim [against him],

E. as it is said, All heave-offering . . . is the priest's (Nu. 5:9).

F. [That is], they [i.e., the priests] have no claim until it is separated as heave-offering.

G. The same holds true (wkn) for a Levite who ate his tithe before its heave-offering was removed (bṭblw):

H. even though he is answerable in [the eyes of] heaven, the priest has no claim [against him],

I. as it is said, that which is taken as heave-offering (Nu. 18:24).

J. You have no claim for it until it is separated as heave-offering.

T. 3:11 (p. 259, l. 32 — p. 260, l. 37)
(A-B: b. Qid. 48b, b. Ned. 84b; C-J: cf. b. Ḥul. 130b)

A farmer is obligated to give approximately two percent of his harvest to the priests as heave-offering and another ten percent to the Levites as first tithe.[11] The question now is whether or not the priests and Levites own their share of the produce even before the farmer has formally designated produce in his crop as heave-offering and first tithe. As we shall see, the produce belongs to the farmer until he specifically designates it as a priestly gift. We have a dispute at A-B followed by a doublet at D-F and G-J.

In A someone eats produce for which the priestly gifts have not yet been designated. Rabbi declares that the user owes the farmer compensation for the entire amount consumed. Since none of the produce was designated as heave-offering or first tithe, it all belongs to the farmer. Yosé disagrees, holding that the appropriate portion of the produce already belongs to the priests and Levites even though it has not been formally separated. Accordingly, the farmer can claim compensation only for that portion of the produce which would remain after these gifts were removed.

The doublet at D-J carries forward Rabbi's view that produce does not become the property of the priests or Levites until it is specifically separated as heave-offering or first tithe. If the farmer eats produce without removing these gifts, therefore, the priests and Levites have no claim against him. He has violated the tithing laws, however, and as in T. 3:9-10, is answerable to heaven.

A. The House of Hillel say, "One separates first tithe from doubtfully tithed produce and removes its [the first tithe's] heave-offering and eats it [the remainder of the first tithe],[12] and does not need to separate second tithe [from the produce at all]."

B. The House of Shammai say, "He [also] must separate second tithe [before eating the produce].

C. "For I say [reading with E: <u>š'ny</u>], 'If second [tithe] is removed, [we assume] first [tithe] is removed, [but] if first [tithe] is removed, [we do] not [assume] second [tithe] is removed.'"

D. And the law is according to the House of Shammai.

T. 3:15 (p. 260, l. 50 — p. 261, l. 53)

E. "[If one] sees him separating (r'h 'wtw š) second tithe — that person is trusted [to have separated] first [tithe]," the words of R. Eliezer.

F. And sages say, "If he is trusted as regards first [tithe], he is trusted as regards second [tithe], [but if he] is trusted as regards second [tithe], he is not trusted [as regards] first [tithe]."

G. "If one separates heave-offering, first tithe and second tithe, and [then] eats of produce —

H. as regards that kind of produce he is trusted [to remove the priestly gifts] but as regards other kinds [of produce] he is not trusted," the words of R. Eliezer.

I. And sages say, "Even in regard to that kind [of produce] he is not trusted,

J. for it is still in his power (l'kl hymnw) to properly tithe for himself and then to go and spoil it [for others]."

T. 3:16 (p. 260, ll. 53-58)
(B: y. Dem. 4:5)

Two separate issues are joined together. In A-D the Houses disagree as to whether or not second tithe need be separated from doubtfully tithed produce. E-J asks whether or not a farmer who is seen separating one priestly gift is trusted to remove others. This latter block of material is linked to the Houses' dispute through the gloss at C. E-J consists of two units: E-F and G-J.

Produce which may already have had tithes removed is subject to the removal of some of the priestly gifts, but not of others. Heave-offering is not removed, since it is assumed that the original owner removed this gift. First tithe, while it is designated, is not given to the Levites unless they can establish a claim to it. This is to insure that all heave-offering due from the crop is separated, since a non-priest who eats heave-offering is subject to death. Heave-offering of the tithe, however, is removed.[13] At issue before us is whether or not the farmer must separate second tithe. The Hillelites rule that he does not need to separate this gift. They assume that the original owner will have separated second tithe, since its separation involves no financial loss on his part (so HD). The Shammaites, however, assume that the original owner separated nothing but heave-offering. They therefore require the householder himself to separate second tithe for the produce. We expect C to supply a reason for the Shammaite opinion, but instead C introduces a new theme and is clearly out of place. We presuppose that both first tithe and second tithe must be removed, in accordance with the Shammaites in B. Under discussion is whether, if one of these gifts is removed, we can assume that the other has been removed as well. This theme is taken up in E-J. Eliezer (E) repeats what has been said in C: one known to separate second tithe is assumed also to separate first tithe. The reason is that the priestly gifts are removed in a set order: heave-offering, first tithe, second tithe.[14] If second tithe is separated, therefore, we can assume the prior gifts, heave-offering and first tithe, also have been separated. Sages do not agree. They hold that if the farmer is willing to separate first tithe, which he must give to the Levites, he will certainly separate second tithe, which he himself eats. It does not follow, however, that one who separates what he himself can eat will also separate that which he must give away.[15]

The consumer in G-I has been observed to separate all the priestly gifts from a certain batch of produce. Eliezer claims that we can now trust him properly to tithe all produce of that kind. The consumer's action, then, is taken to indicate his commitment to the tithing laws for at least this type of produce. Sages hold that we assume nothing. We deem to be properly tithed only that produce from which we have actually observed tithes being removed.

A. One who says, "The second tithe which is in this item is deconsecrated with this <u>issar</u>," but did not specify its [i.e., the second tithe's] location [within the produce] —

B. R. Simeon says, "He has designated [the coin to be consecrated in place of the second tithe]."

C. And sages say, "[He has not consecrated the coin] until he specifies [that the second tithe is] in the northern or southern [portion of the item]."

T. 3:17 (p. 261, ll. 58-60)
(y. Ter. 3:5; cf. M. Ter.
3:5, y. Dem. 7:5)

A. M^cśh b: Rabban Simeon b. Gamaliel and R. Judah and R. Yosé entered the residence of a householder in Achzib. They said, "We do not know how this householder makes ready (mtqyn) his produce [i.e., whether or not his produce is to be deemed properly tithed]."

B. When he sensed [their unease], be brought before them a chest (dlwsqys, cf. Lieberman TK, p. 763) full of gold dinars which he claimed were in the status of second tithe.

C. They said to him, "How do you make ready your produce?"

D. He said to them, "Like this (kk wkk): I say, 'The second tithe which is in this item is deconsecrated with this issar.'"

E. They said to him, "Go and eat your [own] produce. You have gained the coins [i.e., all your coins are unconsecrated], but you have lost your life [because you have not separated second tithe]."

T. 3:18 (p. 261, l. 60 —
262, l. 65)

A farmer wants to designate second tithe in a particular piece of produce and then transfer that status to a coin. The question here is what constitutes proper designation of second tithe such that this status comes to inhere in the produce and so may be removed. Simeon rules that the farmer may make a general declaration that the status of second tithe is in the item. Sages, however, require the farmer to designate as second tithe a particular portion of the item, e.g., its northern or southern half.[16] Only after localizing the status of second tithe in the item may the farmer transfer it to coins. T. 3:18 supports sages' opinion. The householder in Achzib separates second tithe without first localizing it in a specific portion of his produce. His actions are deemed invalid.

A. A haber who dies and leaves produce [the status of which is unknown],

B. even if [he dies] on the same day he brought them in [from the field],

C. lo, they are assumed to be properly tithed.

T. 3:19 (p. 267, ll. 65-66)
(T. Toh. 9:6) (Cf. b. Pes.
4b, 9a; b. A.Z. 41b; b.
Nid. 15b y. Ma. 3:3)

A haber here refers to one who is known to faithfully separate heave-offering and tithes before eating, selling or giving away his produce. If a haber dies, any produce in his possession is assumed to have had heave-offering and tithes removed. This assumption holds even if it appears that he did not have time to properly remove tithes for his harvest.

4:3

A. [If] the householder says, "[This produce is worth] a sela^c^," and someone else says, "A sela^c^," the householder['s bid] has priority, since he adds an extra fifth [of the selling price to what he must pay].

B. [If] the householder says, "A sela^c^," and someone else says, "A sela^c^ and an issar," [the bid] of a sela^c^ and an issar has priority, since it increases the principal (qrn) [i.e., the original purchase price].

C. One who redeems his own [produce in the status of] second tithe adds a fifth [of its selling price],

D. whether [the produce] was [originally his own] or whether it was given to him as a gift.

M. 4:3
(B: b. Arak. 27b; M. Arak.
8:2; C-D: b. Qid. 54b)

M. 4:3-5 considers a new topic — the added fifth. This is a supplementary fee, twenty percent of the selling price of produce, which a farmer pays when transferring the status of second tithe from his own produce to his own coins.[17] If a farmer has four dinars of produce which he wants to replace with his own money, for example, he must consecrate in its place five dinars in coin. This basic law is stated at C-D. The material at A-B supplies a transition to this topic. This it does by introducing the added fifth into M. 4:2's discussion of auctioning off consecrated produce.

In A-B the farmer himself is submitting a bid to establish the selling price of his consecrated produce. He enters the same bid as his competitor. According to A, the farmer's bid is in this case given priority. This is so because he actually pays more for the produce than his fellow bidder, owing the selling price plus an extra twenty percent. B resolves an ambiguity left open by A. It declares that the added fifth is deemed to be a supplementary fee and not part of the selling price itself. The situation in B is such that the second bidder offers a higher market price than does the farmer, but actually will pay less, the farmer owing the one *sela^{c}* he bid plus an extra nineteen *issars* as the added fifth.[18] B rules that in this case the second bidder buys the produce. This is so because he establishes the higher market value of the produce. The extra amount the farmer pays, then, is considered to be a supplement to the market price, not part of the market price itself.

A. *One who redeems his own [produce in the status of] second tithe* [M. 4:3C] —

B. lo, this one is obligated [to pay] the added fifth.

C. The same [law applies to] one who (*'ḥd hkwns*) (1) gathers in [untithed produce] from his own [crop],

D. and to (2) one who buys (*l'ḥd lwqḥ*) and to (3) one who inherits and to (4) one to whom [untithed produce] has been given as a gift,

E. and to (5) one who purchases (*w'ḥd hlwqḥ*) untithed produce from a Gentile or a Samaritan —

F. lo, this one is obligated [to pay] the added fifth [when he himself redeems the second tithe separated from such produce].

T. 4:1 (p. 262, ll. 1-4)

G. How [do we interpret]: "[This one] is obligated [to pay] the added fifth [cf. T. 4:1A]?"

H. He who redeems [the produce] for a *sela^{c}* [i.e., four *dinars*] pays five *dinars* [i.e., twenty-five percent extra].

I. R. Eleazar, the son of R. Simeon, says, "He who redeems [the produce] for a *sela^{c}* pays an extra fifth of a *sela^{c}* [i.e., twenty percent extra]."

T. 4:2a (p. 262, ll. 4-5)

The rule at A-B is carried forward in C-F and is the subject of the gloss at G-I. A-B cites M. 4:3C, declaring that a farmer who transfers the status of second tithe from his produce to his own coins must pay the extra fifth. According to C-F, this is true no matter how the farmer originally acquired the produce from which the second tithe is separated. The point is that the consumer bears ultimate responsibility for seeing that this gift is removed. G-I ask how we are to calculate the added fifth. H declares that the surcharge is to comprise one-quarter of the market price of the produce, that is one-fifth of what the farmer finally pays (H). Eleazar b. Simeon (I) disagrees, arguing that the surcharge is to be equivalent to one-fifth of the market price. It is notable that such a basic question could still be under debate by Ushans.

J. He who redeems [produce in the status of] second tithe —

K. if he wanted to be particular [about the coins he received], let him be particular.

L. How so?

M. [If] he deconsecrates [produce in the status of second tithe] with a <u>sela^{c}</u> and [the <u>sela^{c}</u>] turns out to be of poor quality (<u>r^{c}h</u>),

N. or [if he deconsecrates the produce] with a <u>tressis</u> [i.e., a coin worth three <u>issars</u>] and [the <u>tressis</u>] turned out to be of poor quality,

O. let him exchange [the inferior coin for a coin in better condition].

T. 4:2b (p. 262, ll. 5-7)

A farmer sells produce in the status of second tithe and receives in exchange coins of poor quality. He may, if he so desires, replace these with coins of better quality. This ruling is in line with M. 2:6-9, which declares that the status of second tithe always may be transferred from less desirable to more desirable commodities.

4:4-5

A. They circumvent [the law of] second tithe [so as to avoid paying the added fifth].

B. How so?

C. One says to his adult son or daughter [or] to his Hebrew servant or handmaid,

D. "Take (hylk) these coins and redeem [with them] this [produce in the status of] second tithe [without paying the added fifth]."

E. However ('bl), let him not say this to his minor son or daughter, to his Canaanite servant or handmaid,

F. for their deed is considered to be his deed [ydn kydw].

M. 4:4 (A-D: b. Git. 65a; cf. M. Er. 7:6)

G. [If] he was standing at the threshing floor and had no coins with him, he says to his fellow, "Lo, this produce is given to you as a gift," and then immediately says (ḥwzr w'wmr), "Lo, it is deconsecrated with coins that are at home."

M. 4:5 (b. B.M. 45b)

We have seen that the added fifth is not owed if someone else buys a farmer's consecrated produce or sells produce to him. The present pericopae point out that this is so even if both the coins and the produce remain in the domain of the farmer who designated the produce as second tithe. This view of matters suggests a way in which the farmer can transfer the status of second tithe from his own produce to his own coins without incurring the added fifth. He simply arranges for someone else to buy the produce from him (A-F) or to sell it to him (G). Let us consider each of M.'s cases in more detail.

In the first case, A-F, the farmer directs a member of his household to buy the consecrated produce. Although the produce remains within the householder's domain, the purchase is effected by someone else. The result is that the added fifth need not be paid. E-F emphasize that the family member whom the farmer enlists for the transaction must in fact be able legally to purchase produce. A minor or Canaanite servant (E) cannot engage in commerce on his own.[19] If the farmer has one of these individuals buy his produce, he has not sold the produce to an independent agent. Rather, he has transferred the second tithe to one of his own dependents. It follows that he must pay the added fifth.

G is formally independent of A-F, but adds a second complementary procedure for circumventing the payment of the added fifth. The farmer gives the consecrated produce to his neighbor as a gift. He then declares the status of second tithe to be transferred from it to coins he has at home. He now has consecrated the coins he intends to take to Jerusalem. He does not owe the added fifth, however, since the produce legally belongs to someone else. After the transaction, his friend presumably returns the now-unconsecrated produce to the farmer.

A. They circumvent [the laws of] second tithe to exempt it from the added fifth.

B. How so?

C. One says to his adult son or daughter, to his Hebrew servant or handmaid,

D. "These coins are for you; redeem with them this [produce in the status of second] tithe [M. 4:4A-D]."

E. However, let him not say to him, "Redeem with them on my behalf (ly) this [produce in the status of second] tithe."

F. Said R. Joshua b. Qorḥa, "At first they used to do this. Since swindlers became common (mšrbw hrmyyn), one says to his fellow, 'Lo, this produce is given to you as a gift.' And immediately says to him, 'Lo, it is deconsecrated with (read with G: the coins) I have at home [M. 4:5G],'

G. "provided that he not say to him, 'Lo, it is deconsecrated by the coins I have in my pocket,' until the householder gives [his fellow actual] possession (šyzkm) [of the produce], or until [the householder] rents [to his fellow] his place [in which the produce is kept]."

T. 4:3 (A-D: b. Git. 65a;
F: b. B.M. 45b)

T. explicitly joins M. 4:4 and 4:5, adding glosses at E and G. M. 4:4 has told us that a farmer may avoid paying the added fifth by having another person buy his produce and then give it back to him. It was clear that the other person had to be an autonomous agent, free to conduct business on his own. T. carries forward this notion by declaring that the recipient also must enjoy full control over the coins or produce he is given.

He may not, for example, accept coins on the condition that they be used only to buy the farmer's consecrated produce (E). By accepting coins with this stipulation, the recipient becomes the farmer's agent, and the added fifth will have to be paid.

According to Joshua b. Qorḥa (F), this method was given up when the offspring, or servants, began to accept the farmer's coins, but then refused to use them to buy the consecrated produce. To avoid being cheated, the farmer now gives the other party the consecrated produce, which he then purchases back. Since the person accepting the produce is under the obligation to eat it in Jerusalem, it is to his advantage to allow the farmer to buy it. This tactic is valid only if the second party has legal control over the produce he is given (G). If the recipient does not have legal possession of the produce, it is again as if the farmer purchased his own produce, and therefore he will owe the added fifth. In this way the law of M. 4:5G is linked to the different procedure outlined in M. 4:4.

A. One deposits [produce in the status of] second tithe with a ḥaber [i.e., one who is careful about the laws of cleanness],

B. and [deposits produce separated as second tithe] from doubtfully tithed produce with an ᶜam ha'areṣ [i.e., who is not careful about the laws of cleanness].

C. Aba Ḥilpai b. Qeruya says, "At first they used to do this. They reconsidered and decreed (ḥzrw lwmr) that even with a ḥaber, they do not deposit [produce in the status of second tithe], for you never know who will succeed him."

T. 4:4 (p. 263, ll. 14-16)

The rule at A-B is glossed at C. Produce in the status of second tithe must be kept in a state of cleanness. If the farmer wishes to leave such produce with another person for safekeeping, he therefore must deposit it with a person who is meticulous about the laws of cleanness (A). The stringency is relaxed in regards to produce separated as second tithe from doubtfully tithed produce (B), since such produce may not be in the status of second tithe. Ḥilpai b. Qeruya's gloss at C disagrees with A. Produce in the status of second tithe is not given for safekeeping even to a person who faithfully observes the laws of cleanness. The farmer, who has ultimate responsibility for the produce, cannot be certain that the haber will always be in a position to safeguard the produce.

A. [If the farmer] sets aside (delete with Lieberman: to him) the market price (qrn) [of produce in the status of second tithe] and does not pay the added fifth —
B. R. Eliezer says, "Let him eat [the produce as unconsecrated produce]."
C. And sages say, "Let him not eat [the produce as unconsecrated produce]."
D. Said Rabbi, "The ruling (dbry) of R. Eliezer is preferable (nr'yn) for the Sabbath and the ruling of sages [is preferable] for (add: weekdays)."

T. 4:5 (p. 263, ll. 16-18)
(b. B.M. 44a)

E. One who did not designate heave-offering of the tithe in doubtfully tithed produce —
F. R. Eliezer says, "Let him eat [the produce as properly tithed produce]."
G. And sages say, "Let him not eat [the produce]."
H. Said R. Yosé, "The ruling of R. Eliezer is preferable for the Sabbath and the ruling of sages [is preferable] for weekdays."

T. 4:6 (p. 263, l. 18 —
p. 264, l. 20)

We have a doublet in which different protases, A and E, introduce identical apodoses, B-D and F-H. In each half of the doublet, sages accept a tithing law which Eliezer rejects. T. 4:5 deals with payment of the added fifth. Sages hold the commonly accepted opinion that a farmer who transfers the status of second tithe from his produce to his own coins must pay the added fifth. Eliezer, however, rules that the surcharge need not be paid. He holds, then, that produce in the status of second tithe is deconsecrated in all events by payment of only its market price. T. 4:6 refers to the designation of heave-offering of the tithe in doubtfully tithed produce. Sages declare that this gift must be designated so that the consumer can be sure that all heave-offering has been accounted for in the produce he is about to eat.[19a] Eliezer holds that this gift is not removed from, and therefore not designated in, doubtfully tithed produce. According to his view, produce is designated as heave-offering only if it is known for certain that this gift has not yet been separated from the batch.

Rabbi and Yosé, D and H, attempt to reconcile the opposing views of sages and Eliezer. The sages' view defines the norm, requiring the added fifth to be paid in the case of second tithe and requiring a formal declaration for produce separated as heave-offering of the tithe. Eliezer's ruling is taken to apply to the Sabbath, when the householder is prohibited from deconsecrating his produce (cf. M. Beṣ. 5:2). We would normally expect that the produce, the deconsecration of which is only partially completed, may not be eaten on the Sabbath but must be kept until a weekday when it can properly be deconsecrated. Eliezer's ruling, as read by Yosé, allows the householder to eat such produce on the Sabbath with the understanding that he will complete its deconsecration when the Sabbath is over.

A. Two brothers, two partners [or] a father and son may redeem [produce in the status of] second tithe for each other [so as to avoid payment of the added fifth], and they give each other poorman's tithe [cf. M. 4:4-5].

B. Said R. Judah, "May a curse come upon him who gives poorman's tithe to his father."

C. They said to him, "What if both of them [i.e., father and son] were poor?

D. One's wife redeems on his behalf [produce in the status of second] tithe [without paying the added fifth] according to the words of R. Simeon b. Eleazar.

E. One's wife does not redeem on his behalf [produce in the status of second] tithe.

F. All those whom they said redeem [produce in the status of] second tithe for each other [so as to avoid paying the added fifth], give each other poorman's tithe [and since a wife is not eligible to receive poorman's tithe from her husband, it follows that she may not purchase second tithe from him].

G. R. Simeon b. Eleazar concedes that as regards one's wife, even though she does redeem [produce in the status of] second tithe [on her husband's behalf], she is not given poorman's tithe.

T. 4:7 (p. 264, ll. 20-26)
(A-C: y. Peah. 5:4, cf. b. Qid. 32a; D-E: y. Er. 7:6, y. Qid. 1:3; y. M.S. 4:4, b. Qid. 24a)

The rule at A is glossed at B-C. There follows a dispute at D-E which bears a secondary expansion at F and G. Any legal adult who purchases a farmer's produce in the status of second tithe does not pay the added fifth (M. 4:4). A links such people with those who are eligible to receive poorman's tithe from the farmer, that is, people who are financially independent of the farmer.[20] The question now is whether or not a wife, who is clearly a dependent, is nonetheless considered to be a legal adult such that she may purchase second tithe from her husband without paying the added fifth. Simeon b. Eleazar (D) declares that since the wife is an adult, she may purchase second tithe from her husband without his paying the added fifth. E differs, declaring that the wife does not exempt her husband from the added fifth. As a dependent, she is deemed to be in the same category as a minor or a Canaanite slave. F supports E's ruling by reminding us that these persons who are the farmer's dependents and ineligible to receive poorman's tithe must pay the added fifth when purchasing the farmer's second tithe. Since a wife is ineligible to receive poorman's tithe from her husband, it follows that she may not purchase second tithe from him unless she also pays the added fifth. Simeon b. Eleazar responds, in G, by claiming that the two statuses of adult and dependent are not linked. The wife is a legal adult as regards second tithe, while at the same time she is a dependent as regards her ineligibility to receive poorman's tithe from her husband.

A. They separate heave-offering of the tithe for produce [owned by] an <u>ᶜam ha'areṣ</u>
B. even though this leads to a misdeed [on the part] of those who follow [<u>šᶜwśh tqlh lb'ym 'ḥryw</u>] [i.e., those who will later buy the produce].
C. How so?
D. [If the <u>ᶜam ha'areṣ</u> brings the produce out of his house and says to [the <u>ḥaber</u>], "Separate [heave-offering of the tithe] for this,"
E. he separates [heave-offering of the tithe] only for the amount he buys [<u>ᶜd mqwm šhw' lwqḥ</u>].

T. 4:8 (p. 264, ll. 26-28)

F. They deconsecrate [produce in the status of] second tithe [by transferring its status to] coins [belonging to an] ᶜam ha'areṣ

G. even though this leads to a misdeed [on the part] of those who follow.

H. How so?

I. [If the ᶜam ha'areṣ] brings the coins out of his house and says to [the ḥaber], "Deconsecrate [the produce] with these,"

J. he [the ḥaber] deconsecrates only [the amount of produce] that he intends to buy.

K. And he [the ḥaber] immediately gives him [the ᶜam ha'areṣ] change [pwrṭ] for a dinar [i.e., he trusts the ᶜam ha'areṣ not to give him a coin he has just consecrated as second tithe].

L. If [the ᶜam ha'areṣ gave him the coin] on account of a debt, it is permitted [for the ḥaber to accept the coin since the ᶜam ha'areṣ would not give him a coin in the status of second tithe].

M. If [the ᶜam ha'areṣ gave him the coin] as a loan, lo, it is forbidden [for the ḥaber to accept the coin because the ᶜam ha'areṣ might not know that it is forbidden to loan coins in the status of second tithe],

N. until he [the ᶜam ha'areṣ] goes into his house and returns [with a different coin].

T. 4:9a (p. 264, ll. 28-32)
(F: cf. M.T.Y. 4:5)

The pericope consists of a doublet at A-E and F-J, followed by appended material at K-N. A ḥaber is purchasing produce from an ᶜam ha'areṣ. Since such produce is of doubtful status, heave-offering of the tithe and second tithe must be removed (cf. T. 3:15). The ḥaber wants to separate these gifts before making the purchase so that the consecrated produce will remain with the ᶜam ha'areṣ. To do so creates a problem, however, in that others who purchase produce from the ᶜam ha'areṣ will assume these gifts have not been removed and so will attempt to separate them again. If the ḥaber already has removed heave-offering of the tithe and second tithe, that which the later purchaser sets aside will remain unconsecrated.[21] In order to avoid this, C-E and H-J limit what the ḥaber may tithe by declaring that he may separate heave-offering of the tithe or second tithe only for that produce he himself will buy. The remainder of the ᶜam ha'areṣ's produce thus remains in its doubtful status and its purchaser properly removes heave-offering of the tithe and second tithe from it.[22]

K-N discuss a case in which an ᶜam ha'areṣ has coins in the status of second tithe as a result of a transaction such as that described in F-J. The question is whether or not we trust the ᶜam ha'areṣ to reserve his consecrated coins for their proper purpose. In L the ᶜam ha'areṣ gives the ḥaber a coin in order to receive change or to repay a debt. Although the ḥaber is not certain of the coin's status, he may accept it on the assumption that the ᶜam ha'areṣ knows not to use consecrated coins for secular purposes. The haber may not accept a coin as a loan, however, since the ᶜam ha'areṣ may loan a consecrated coin, reasoning that it will soon be returned. If the ᶜam ha'areṣ goes into his house to get a coin, however, we assume he has done so in order to bring out an unconsecrated coin, and the ḥaber may therefore accept the coin even as a loan.

O. A minor (qṭn; Lieberman emends to sṭn: wholesaler) who is selling [produce] in the market —

P. they use him to separate second tithe (mpryšym ᶜlyw) [so as to avoid paying the added fifth].

Q. How so?

R. He [i.e., the farmer] gives him [i.e., the minor] a tressis,

S. whether it is in the hand of the purchaser [i.e., the minor] or the hand of the seller [i.e., the farmer] or [already] dropped into the money-purse (b' b'wnqly wnpl; reading with Lieberman b'ynply),

T. he [i.e., the farmer] designates [second tithe in the produce and transfers to the coin] [without paying the added fifth], and may continue to do so (whwlk).

U. [If the farmer] receives a dinar from [the minor], he designates [second tithe in his produce and sells it to the minor] and may continue to do so.

V. Moreover (wl' ᶜwd 'l'), one may bring a dinar from his house, separate second tithe with it [i.e., by declaring second tithe in the produce to be transferred to the coin], leave, come back and finally give the coin [to the minor] and say to him, "This is the coin with which I deconsecrated [my produce]," [and since the minor can now go through the motions of buying the produce, the added fifth is not paid].

T. 4:9b (p. 264, ll. 32 — p. 265, l. 36)

T. liberalizes the rules for selling second tithe to another person so as to avoid payment of the added fifth. As we saw in M. 4:4, a farmer can avoid paying the added fifth by giving coins to a legal adult, who then goes through the motions of purchasing with them the farmer's second tithe. O-P declare that the transaction is valid even if the buyer is a minor, provided that the minor is functioning as an independent agent in the market.[23] The remaining three rules, R-T, U and V, indicate that the transaction is deemed valid, exempting the farmer from the added fifth, even if it is improperly carried out. In R-T the farmer has not given the potential buyer full control over the coins before he declares the produce sold for those coins.[24] In U the farmer receives payment for his second tithe before this gift has actually been separated from the produce. Finally, the farmer, in V, transfers the status of second tithe to the coin before giving the coin to the prospective buyer. If the coin eventually is given to the buyer and the buyer goes through the motions of buying the consecrated produce, the added fifth is no longer owed.[25]

A. (1) [If a ḥaber] receives from him [i.e., an ^{c}am ha'areṣ] ten gold dinars [with which to buy second tithe belonging to the ^{c}am ha'areṣ],
B. he should not use all of them to deconsecrate [the ^{c}am ha'areṣ's produce] but [he should deconsecrate only as much produce as can be deconsecrated] with one of [the coins].
C. (2) [If he] bought from him ten kors of wheat,
D. he should not separate [second tithe] for the entire lot, but only for one of [the kors].
E. (3) [If he] bought from him ten kegs of wine,
F. he should not separate [second tithe] for the entire stock, but only for one of [the kegs].
G. (4) [If he] bought from him ten bundles of greens,
H. he should not separate [second tithe] for the entire load (h'nqly), but only for one of [the bundles],
I. so as not to cheapen (reading with E, ed. princ: (hpḥyt) their value in his estimation (b^{c}ynyw) [by flooding the

market with produce in the status of second tithe which the ^{c}am ha'areṣ is anxious to sell].

T. 4:10 (p. 265, ll. 36-40)

T. carries forward the theme of T. 4:9a. A ḥaber intends to purchase produce from an ^{c}am ha'areṣ. He wants to separate second tithe for the produce before the purchase, however, so that the second tithe will remain with the ^{c}am ha'areṣ. T. declares that the ḥaber may not leave the ^{c}am ha'areṣ with a large amount of second tithe. The reason is given in I. The ^{c}am ha'areṣ is not concerned with the priestly gifts and so will be willing to sell his large batch of second tithe at a low price. He thereby decreases the value of the second tithe.

A. One sells [produce in the status of second] tithe as [lšm] (add with A: unconsecrated) produce.
B. Let him not sell it [i.e., produce in the status of second tithe] as [second] tithe [because the purchaser knows the farmer is anxious to sell it] [cf. T. 4:10I].
C. [If] he has unconsecrated [produce] and [produce in the status of] second tithe mixed together and wants to sell them,
D. he sells first [an appropriate amount of produce either as unconsecrated or as second tithe and then sells the rest of the produce as the other type] (r'šwn r'šwn) and puts the coins [for each] in a [separate] money-box.
E. If he sold them all in one lump (ntnn kwln bkrk), he receives their payment in one lump.
F. Whatever he wants [to sell as unconsecrated produce] he sells as unconsecrated produce and whatever he wants [to sell as second tithe] he sells as second tithe.

T. 4:11 (p. 265, l. 40 —
p. 266, l. 44) (Cf. y.
M.S. 4:1; E-F: y. A.Z. 5:1)

The pericope before us is composed of two rules: A-B and C-F. A farmer is selling produce in the status of second tithe in order to receive coins to take to Jerusalem. According to A-B, he should not announce the fact that he is selling consecrated produce. If the purchasers know that the farmer is selling second tithe, they will assume he is anxious to be rid of it, and they therefore will offer a lower price (HD). The result will be a reduction in the amount of currency the farmer takes to Jerusalem.

C-F concerns a case in which a farmer has consecrated and unconsecrated produce which became mixed together. Although the farmer does not know which part of the mixture is consecrated, he does know the proportion of consecrated produce in the whole. The farmer wants to sell all of the produce, setting aside coins equal in value to the consecrated produce, since he will have to take these coins to Jerusalem. He may do this in one of two ways. He may sell an appropriate amount of produce as second tithe, putting the receipts in one coin-box. He then sells the remainder of the produce as unconsecrated merchandise, putting the receipts in another box. Alternatively, the farmer may sell the mixture all at once. He must then set aside as second tithe an appropriate value of coins from the receipts.

4:6

A. [If a purchaser] took possession (mšk) from him [i.e., a farmer] [of produce in the status of second] tithe [which is worth] a selaᶜ,
B. and did not have time (hśpyq) to redeem it [i.e., pay for it] before [the price] went up to two [selaᶜs],
C. he pays him [i.e., the farmer] one selaᶜ,
D. and earns a profit [on the produce he receives] of one selaᶜ,
E. and [one selaᶜ's worth of produce he acquires is] his [in the status of] second tithe.
F. [If a purchaser] took possession from him [i.e., a farmer] of [produce in the status of second] tithe [which at the time of acquisition is worth] two selaᶜs,
G. and did not have time to redeem it before [the price] went down to one selaᶜ,
H. he pays him one selaᶜ in unconsecrated coin and one selaᶜ in [one of] his own coins [in the status of] second tithe.
I. If he [i.e., the farmer] was an ᶜam ha'areṣ, he [i.e., the purchaser] gives him [coins in the status of] doubtful [second tithe].

M. 4:6
(A-C: b. Qid. 54b)

The selling price of produce is finalized when the purchaser takes possession of it, as A-C make clear. This fact creates a problem peculiar to the deconsecration of produce in the status of second tithe. As we recall (M. 4:1), the amount of money necessary to deconsecrate such produce depends on the market value of the produce at the time payment is made. In the present case, this market value changes between the time the purchaser takes possession of the produce and the time he is ready to pay for it. The amount of money he has agreed to pay, therefore, is not equivalent to the amount of money needed to deconsecrate the produce. At issue is the quantity of produce which is deconsecrated on account of the sale. As we shall see, the amount of produce deconsecrated with a set amount of money depends on the value of the produce when payment is actually made, regardless of the agreed upon purchase price. Formally, we have a doublet formed of the contrasting cases at A-E and F-H. I qualifies F-H.

At A-E the amount of money the purchaser has agreed to pay is insufficient fully to deconsecrate the produce. He has agreed to pay one *sela*^{c} for produce which now is worth two *sela*^{c}*s*. While the purchaser indeed acquires the two *sela*^{c}*s'* worth of produce for the one *sela*^{c}, as has been agreed upon, the sale effects the deconsecration of only half of the produce, that is one *sela*^{c}*'s* worth. Since no coins were given in exchange for the second *sela*^{c}*'s* worth of produce, it retains its consecrated status (E).[26] The purchaser therefore must be sure to use it as second tithe.

The reverse is discussed in F-H. The purchaser has agreed to buy for two *sela*^{c}*s* produce which later is worth only one *sela*^{c}. Since the produce is worth only one *sela*^{c}, only one of the two *sela*^{c}*s* which is paid to the merchant becomes consecrated in place of the produce. The second *sela*^{c} which is paid clearly remains unconsecrated. This creates a problem for the merchant since there is no way for him to determine which one of the two *sela*^{c}*s* he receives is consecrated on account of the sale and which one is unconsecrated. There is a danger that he will mistakenly treat the consecrated coin as though it were unconsecrated. H therefore requires the buyer to pay over one *sela*^{c} of the purchase price in already-consecrated coin. In this way both coins left with the merchant are consecrated and possible confusion is avoided.

I repeats a common theme. If the purchaser is an ^{c}am ha'areṣ, he is not given an object certainly in the status of second tithe. This is so because he is not trusted properly to dispose of such things.

J. [If he] redeemed [i.e., paid for] from him [a merchant] produce in the status of second] tithe [which was worth] a sela^{c}
K. and did not have time to take possession of it (lmškw) before [its price] went up to two [sela^{c}s] [M. M.S. 4:6A-B],
L. that which he redeemed [paid for] he has [validly] redeemed
M. and the law has been satisfied (mdt hdyn bynyhm).
N. [If he] redeemed [paid for] from him [produce in the status of second] tithe [which at the time of purchase is worth] two sela^{c}s
O. and did not have time to take possession of it before [the price] settled at one sela^{c} [M. M.S. 4:6D-E],
P. that which he redeemed [paid for] is redeemed
Q. and the law has been satisfied,
R. for [as regards produce in the status of] second tithe, the act of redeeming it [takes place at the moment of] its sale.
S. Rabban Simeon b. Gamaliel and R. Ishmael in the name of R. Yoḥanan b. Baroqa said, "Also: [as regards produce] dedicated [to the Temple, its redemption [takes place at moment of] its acquisition [by the purchaser]."

T. 4:14c (p. 267, ll. 62-66)
(T. Arak. 4:4; S: y. M.S. 4:6, y. M.S. 5:1)

T. consists of a doublet at J-M and N-Q. S is an autonomous unit of law set as a gloss to R. We have here the opposite of M.'s case. A purchaser pays for produce in the status of second tithe, intending to take possession of it later. Since the price already has been paid, the produce is deemed to be fully deconsecrated, and subsequent changes in value are of no concern, as R explains. S declares that in the case of dedicated items, deconsecration occurs at the time the purchaser takes possession of the item, not the time at which payment is made. I do not know why there should be such a difference between items in the status of second tithe and items dedicated to the Temple.

4:7

A. One who redeems [i.e., buys produce in the status of] second tithe but does not make a declaration [that he is redeeming the produce] —

B. R. Yosé says, "It is sufficient for him [simply to pay for the produce without making a declaration]."

C. R. Judah says, "He must make it explicit."

D. [If] he was speaking to his wife about matters relevant to her divorce contract or her bride-price and did not make it explicit —

E. R. Yosé, "It is sufficient for him [simply to give her the contract or bride-price without a declaration]."

F. R. Judah says, "He must make it explicit."

M. 4:7
(D-F: b. Qid. 6:1, y. M.S. 6:1; cf. T. Qid. 2:8, T. Git. 8:8)

The dispute between Yosé and Judah is whether a physical act is sufficient to effect a change of status or whether the act must be accompanied by an oral declaration. We have two disputes which investigate this question, only the first of which is relevant to our tractate. In A-C a farmer buys consecrated produce intending to transfer its status to coins. He does not, however, make his intention known to the merchant. Yosé holds that the mere fact of buying the produce is sufficient to deconsecrate it. Judah disagrees. He holds that the transfer of the status of second tithe occurs only if the purchaser makes known his intention to do so. D-F explore a more ambiguous case. A householder is involved in a discussion with a woman concerning marrying or divorcing her. In the middle of the conversation, he hands her a divorce contract or her bride-price.[27] In this case, the householder's intention is known even though no formal declaration to that effect is made. Yosé (E) of course will consider the transaction valid since in any case he deems the physical act alone to be probative. Judah's opinion is of interest in that he holds that even in this case, the transaction has no effect. That is, for Judah even general knowledge of the householder's intent is not sufficient. He must make a specific declaration of his intention when he hands over the divorce contract or bride-price in order to be divorced or married.[28]

A. [If] one was traveling from place to place and had with him coins [in the status of second tithe which he wants to spend on produce to be eaten as second tithe in Jerusalem] —
B. If he plans [eventually] (^{c}<u>tyd</u>) to return to the place [from which he set out],
C. he deconsecrates [i.e., spends] [the coins] at the value [they have] in the place [from which he set out].
D. But if not [i.e., if he will not return to the place from which he set out], he deconsecrates [the coins] at their value in the local market.

T. 4:14a (p. 267, ll. 57-59)

The issue, as in M. 4:8, is to determine the proper value at which coins in the status of second tithe are spent. We now have a householder who has acquired coins in one area but wishes to spend them in another area, where they have a different purchasing power. According to A-C, the farmer spends the coins according to the value they have at his permanent residence. This is the value he himself deems the coins to have. If he does not expect to return to his former residence (D), he spends the coins at the value in the local market, since he has no other point of reference. The assumption here, familiar from M., is that the value of coins is established by market conditions.

E. [If he] separated heave-offering, first tithe or second tithe but did not designate [the separated produce to be consecrated] —
F. R. Yosé says, "It has become consecrated."
G. R. Judah says, "It has not become consecrated [cf. M. M.S. 4:7A-C]."
H. Rabbi says, "If he plans to make the designation later (^{c}<u>tyd lqrwt</u>), [the status of] consecration is not in [the produce] until he makes the declaration.
I. "If he does not intend to make the declaration later, it has become consecrated.

T. 4:14b (p. 267, ll. 59-62)
(T. Ter. 3:3)

E-G restate the dispute of M. M.S. 4:7. In H-I Rabbi reconciles the two opposing opinions. The act of separation alone is sufficient, according to Rabbi, unless the farmer intends to make a formal declaration at some later time. In such a case, the farmer does not intend to have the separated produce become consecrated at the time of separation, and we do not therefore consider it to be consecrated until he has made his declaration. If the farmer does not plan to make a declaration at all, his act of separation is sufficient to consecrate the produce, as Yosé has ruled.

4:8

A. One who sets aside an issar [in the status of second tithe and takes it to Jerusalem] and ate [as second tithe produce purchased] against (^{c}lyw) half of its value,

B. and [then] went to another area [in Jerusalem] and lo, [an issar] is worth a pundium [that is, twice its previous value so that the money remaining is worth a full issar of produce] —

C. [he] eats [against its value as second tithe produce worth] another issar.

D. One who sets aside a pundium [in the status of second tithe] and ate [as second tithe produce purchased] with half of its value,

E. and [then] went to another area and, lo, [a pundium] is worth an issar [that is, half of its previous value so that the money remaining is worth only a quarter of a pundium or half an issar] —

F. [he] eats [against its value as second tithe produce worth] another half [issar].

G. One who sets aside an issar [in the status of second tithe] eats against its account [in Jerusalem an amount such that no more than] one-eleventh of an issar's value [remains if it is of doubtful status] or so that no more than one-hundredth of the issar's worth [remains if it is surely in the status of second tithe].[29]

H. The House of Shammai say, "[In] all [cases, whether the coin is doubtfully in the status of second tithe or certainly in that status, the farmer eats against its account an amount such that no more than] one-tenth [of an issar's value remains]."

I. And the House of Hillel say, "[If the issar is] certainly [in the status of second tithe, no more than] one-eleventh [of its value may remain] and [if it is] doubtfully [in the status of second tithe, no more than] one-tenth [of its value may remain]."

M. 4:8
(A-B: y. Qid. 1:1)

This pericope is a triplet (A-C, D-F, G-I) which continues the concern of M. 4:6 in determining the value of coins in a fluctuating market. A farmer sets aside a coin in the status of second tithe intending to deconsecrate it by eating as second tithe in Jerusalem produce of equivalent value. After the farmer uses up part of the coin's value, however, the market-price of produce shifts. We want to know at what market-price the remaining value of the coin is spent. A-C and D-F state that the remaining value of the coin is spent at prevailing market prices, a principle familiar from M. 4:1. G-I introduce a related but distinct theme. Its point apparently is that consecrated money may be deemed fully deconsecrated even if a small amount of its value remains unspent. The material is difficult to interpret, however, because of the eliptical nature of its language. It is necessary, therefore, to explain how I arrive at my conclusion.

Before turning to the substance of G-I, we need to consider this unit's formal structure. We have two blocks of material, the statement at G plus the appended Houses' opinions at H-I, giving us three opinions in all. G is unintelligible as it stands, since we do not know to what its two figures refer. We must read back into G the distinction made in H-I between coins certainly in the status of second tithe and coins doubtfully in that status. The point of the entire discussion, however, is still unclear. This unit does take on meaning, however, in light of the material presented in the first half of the pericope. As we have seen, A-F introduce the notion that a farmer need not spend at one time the full value of his consecrated coins. That is, he may spend part of the coin's value, allowing the remaining value of the coin to retain its consecrated status. The point of G-I seems to be that if the remaining value is of an insignificant amount, it may be ignored, so that the coin is deemed fully deconsecrated. Under discussion in G-I, accordingly, is the value of second tithe which may be ignored in this way. It is clear from I that one may ignore a larger portion of the coin's value if the coin is of doubtful status than if the coin is surely in the status of second tithe. It is not clear to me, however, how the exact figures supplied in G-I are to be explained.

Commentators to this pericope have offered a number of diverse explanations, all of which are unconvincing. Maimonides holds that the unit is intelligible if we assume that the farmer must pay an added fifth when purchasing produce with his consecrated coins. In this way, one-tenth of an issar becomes that amount of a coin's value with which a farmer can purchase one-eighth of an issar's worth of produce. One-eighth of an issar is equivalent to one peruṭah, the smallest amount of money of legal significance (cf. M. B.M. 4:7). One-eleventh of an issar (H) therefore represents an amount of money that will purchase less than a peruṭah's worth of produce. Although Maimonides can account for most of the figures before us, he offers no evidence that the added fifth must indeed be paid in these cases. Without this crucial point established, his interpretation is without basis.

TYY argues that the issue is the amount of produce freed from the obligation of second tithe when the farmer sets aside an issar in coin as second tithe for the produce. The Shammaites, according to TYY, allow the farmer to eat ten issars' worth: that is, nine issars' worth of unconsecrated produce and one issar of produce which was designated as second tithe and deconsecrated with the coins the farmer set aside. While this explanation does make sense of the Shammaite opinion, it does not account for the figures which occur in the Hillelite opinion or in the opinion of G. H furthermore requires that we treat this unit independently of its redactional context.

Albeck holds that the dispute concerns whether or not the farmer must eat as second tithe in Jerusalem produce worth more than the value of the coin that was set aside. According to Albeck, the figures in the Houses' dispute are to be interpreted as the number of tenths of a coin's value which must be eaten by the farmer in Jerusalem. The Shammaites, then, hold that the produce must be of value equivalent to the value of the coin, that is, ten tenths. For produce certainly in the status of second tithe, the Hillelites hold that the farmer must eat an extra amount, eleven-tenths in all. This is to insure that the full value of the coin is consumed even if the purchasing power of the coin rises before the farmer purchases produce in Jerusalem. While Albeck's interpretation is sensitive to the thematic context of the pericope, it requires us to read the issue of tenths into the text. The Hebrew text, however, does not support such a reading.

A. One who sets aside an issar [in the status of second tithe and takes it to Jerusalem] and eats [as second tithe produce purchased] against half of its value
B. and goes to another area and, lo, [an issar] is worth a pundium [that is, twice its previous value]
C. eats (w'kl) [as second tithe produce worth] another issar.
D. One who sets aside a pundium [in the status of second tithe] and eats [as second tithe produce purchased] against half of its value
E. and goes to another area and, lo, [a pundium] is worth an issar [that is, half its previous value]
F. eats (w'kl) [as second tithe produce worth] another half [issar] [= M. M.S. 4:8A-F].
G. This is the general [rule]:
H. Before he has [fully] redeemed it [i.e., before he has deconsecrated the produce by paying for it], any gain in value accrues proportionately [to the consecrated produce and the consecrated coins] and any loss in value accrues proportionately [to the consecrated produce and the consecrated coins].
I. After he has redeemed it [i.e., deconsecrated the produce by paying for it fully], any gain in value accrues [entirely] to the [coins consecrated as] second [tithe] and any loss in value accrues [entirely] to the [coins consecrated as] second [tithe].

T. 4:15 (p. 267, l. 66 — 268, l. 71)

T. cites M. 4:8 and links it to the underlying principle of M. 4:6. The rule at G-I takes as its subject produce in the status of second tithe in contrast to A-F which deals with coins in that status. The same principle emerges, however. Once a purchaser pays for produce in the status of second tithe, the sale is completed, and the status of second tithe is deemed to be transferred to the coins.

4:9

A. All coins that are found (some mss. add: in Jerusalem)[30] [and the status of which are unknown],

B. lo, these [are deemed to be] unconsecrated [i.e., not in the status of second tithe],

C. even if [one finds] gold dinars [mixed in] with silver and with [copper] coins [and suspects that this collection was set aside for a special purpose].

D. [If] he found among them [i.e., in a batch of coins] a potsherd upon which was inscribed [the word] "tithe" (m^cśr)

E. lo, this [batch of coins] is [in the status of second] tithe.

M. 4:9
(Cf. M. Sheq. 7:1-2)

The point emerges from the contrast between A-C and D-E. Coins, the status of which is in doubt, are deemed to be unconsecrated unless there is clear evidence to the contrary (D-E). Items are deemed to be in a consecrated status, then, only if there is good reason to assume so.

4:10-11

A. One who finds a vessel upon which is inscribed [the word] "offering" (qrbn) —

B. R. Judah says, "If [the vessel is made] of clay, it [the vessel] is unconsecrated, but what is in it is [in the status of] an offering.

C. "If [the vessel is made] of metal, it is an offering, but what is in it is unconsecrated."

D. They said to him, "It is not usual for people to put unconsecrated goods into [a container which is in the status of] an offering."

M. 4:10

E. One who finds a vessel upon which is inscribed [the letter]
(1) "qof," [the produce it contains is in the status of] an offering (qrbn),
(2) "mem," [the produce it contains is in the status of tithe (m^cśr),

(3) "dalet," [the produce it contains is] doubtfully tithed (dm^{c}y)

(4) "ṭeṭ," [the produce it contains is] certainly untithed (ṭbl)

(5) "taw," [the produce it contains is in the status of] heave-offering (trwmh),

F. for in the time of danger they wrote [only the letter] "taw" instead of [writing out the full word] "trwmh" (heave-offering).

G. R. Yosé says, "All [of the letters stand for] the names of individuals [and therefore are not taken to denote the status of produce in the vessel]."

H. Said R. Yosé, "Even if one found a cask full of produce and on [the cask] was inscribed [the word] 'heave-offering,' lo, the (pieces of fruit in it) are unconsecrated.

I. "For I say that last year it was filled with produce [in the status of] heave-offering but [subsequently] it was emptied [and refilled with other produce]."

M. 4:11
(b. Yeb. 115b)

M. 4:10-11 carry forward the theme of M. 4:9 concerning found items the status of which is in doubt. Someone finds produce in a jug which bears an inscription denoting a consecrated status. A-D and E-G ask how we interpret this inscription if it is ambiguous, either because it can apply to the vessel or to the produce it contains (A-D) or because it is abbreviated (E-G). Yosé (G and H-I) holds that inscriptions are of no probative value at all. His view thus stands in opposition to the entire discussion.

Judah considers a case in which a jug of produce is inscribed as an offering to the Temple (A-D). The inscription can be taken as referring to the jug itself or to the produce or to both. According to Judah, we can determine whether the inscription refers to the jug or the produce on the basis of the jug's composition. If the jug is made of clay, he rules that the jug is unconsecrated, and the label therefore must refer to the produce it contains. Presumably, clay vessels were of too low a value normally to be dedicated to the Temple. If the vessel is of metal, and therefore of higher value, we deem the inscription to refer to the vessel itself. Judah concludes that in this case, since the inscription refers to the container, the produce inside may be deemed unconsecrated. D disagrees, claiming that if we assume the vessel is consecrated, we must make the same assumption of its contents.

E-G consider another possible source of doubt concerning the status of produce one finds. This material is linked to the foregoing by the superscription at E. We have a list of common abbreviations used to mark vessels containing food in various consecrated statuses. F, which specifically glosses only E(5), explains that the owner used abbreviations rather than full words in order to save time. This gloss sets the stage for Yosé's disagreement in G. As stated above, he holds that inscriptions may not be used to determine the status of produce. He thus rejects the meanings E imputes to these symbols. His view is carried forward, in a different context, in H-I. Yosé now rules that in all events notations may not be used for identifying the status of a vessel's contents.

A. [If] one found a vessel upon which was inscribed [the letters] "aleph," "dalet," "resh," [or] "taw,"

B. lo, it [the produce in the jug] is [in the status of] heave-offering.

C. [If the vessel has inscribed on it the letters] "yod" [or] "mem,"

D. lo, it [the produce] is [in the status of first] tithe.

E. [If it was inscribed with the letters] "pe" [or] "shin,"

F. lo, it is [in the status of] second [tithe].

G. But the sages say, "All of these [letters denote] the names of individuals [M. 4:11G].

H. "But [if] one found a cask full of produce and it had inscribed on it [the word] 'heave-offering,'

I. "lo, [the produce in it] is [in the status of] heave-offering,

J. "[and if the cask was inscribed with the word] 'tithe,'

K. "lo, [the produce in it] is [in the status of first] tithe."

L. Said R. Yosé, "Even if one found a cask full of produce and it was inscribed with [the word] 'heave-offering,'

M. "lo, these [pieces of fruit in the cask] are unconsecrated.

N. "For I say that last year it was filled with [consecrated] produce but [someone] [subsequently] emptied [it and refilled it with unconsecrated produce [M. 4:11H-I].

O. "If [the cask] is new [that is, not left over from the previous year], I say he filled it [produce in the status of] heave-offering but then emptied it and refilled it with unconsecrated produce."

P. R. Yosé concedes that if [the word] is inscribed on a shard and placed on the cask, or on [a piece of] papyrus and placed over [the vessel's] opening —

Q. [if it says] "heave-offering," lo, it is [in the status of] heave-offering,

R. [if it says] "tithe," lo, it is [in the status of first] tithe.

T. 5:1 (p. 268, ll. 1-8)

T. continues M. 4:11's discussion of how one uses inscriptions on jugs to determine the status of produce contained therein. The anonymous view in A-F declares that inscriptions which might denote a consecrated status are taken as evidence that the produce in question is consecrated. As in M. 4:10E-F, the point is that if there is reason to believe the produce is consecrated, we must treat it as such.[31] Sages, G-K, reject this view, holding that only inscriptions which unmistakably refer to a consecrated status are used as evidence for determining the status of the produce. Yosé, L-N, argues a third point. He claims that in no case is an inscription to be used as evidence for concluding that a batch of produce is consecrated. That is, as in M. 4:11G-I, produce is deemed to be unconsecrated unless there is clear evidence to the contrary. O-R consider two possible objections to Yosé's view. In O the vessel is new, so that the label cannot be assumed to reflect conditions of the previous year. The label is ignored nonetheless. In P-R the label was laid on top of the jug in such a way that it could be removed easily. In this case, Yosé concedes that the fact that the label is still in place is compelling evidence that the produce in the jug is consecrated.

A. "<u>[If] he found a clay vessel on which is inscribed [the letter] 'qof' [or the word] 'offering' (qrbn)</u> [= M. 4:10A],

B. "it [the vessel] is [in the status of] an offering and what is in it is [in the status of] an offering," the words of R. Meir.

C. But sages say, "It is not usual for people to sanctify a clay vessel [as an offering]."

D. [If] he found a metal vessel on which is inscribed [the word] "offering" (qrbn) —

E. if it is empty, it is forbidden to use it until they know that [although] it was [once in the status of] dedication, it has been redeemed [i.e., deconsecrated].

T. 5:2 (p. 268, l. 8 — 269, l. 11)

F. [If] he found a clay vessel on which is inscribed [the word] "offering" [M. 4:10A] —

G. lo, [the produce is in the status of] an offering.

H. [If there was inscribed on the vessel the words] "tithes for the [holy] city" (m^{c}śr ^{c}yr) [that is, tithes to be brought to Jerusalem] —

I. lo, it is in the status of second] tithe [to be taken to the holy] city.

J. [If the vessel has inscribed on it the words] "for Joseph" (lšm ywsp) or "for Simeon" (lšm šm^{c}wn) —

K. this says nothing [about the status of the produce in the vessel].

L. [If the vessel has inscribed on it the words] "to be brought up to Jerusalem to be eaten," —

M. lo, these [pieces of fruit] are unconsecrated.

N. [If the vessel has inscribed on it the word] "tithe,"

O. they subject it [i.e., the produce] (mṭylyn ^{c}lyw) to the restrictions [applicable to] both [first tithe and second tithe since it is not clear to which status the inscription refers].

T. 5:3 (p. 269, ll. 11-14)

A-F go over the ground of M. 4:10A-D. The question is whether or not an inscription denoting a consecrated status refers to the status of the jug on which it is placed or to the status of the contents of the jug. G-P consider the implication of a number of diverse inscriptions for determining the status of produce found in a jug.

Meir, A-C, declares that a clay vessel bearing the inscription "offering" is deemed to be in a consecrated status. In contrast to Judah in M. 4:10, he holds that an inscription denoting a consecrated status is always taken to refer to the jug itself. It is, after all, the jug that is marked. Judah's view, that the inscription is assumed only to apply to jugs made of metal, is ascribed to sages in D-F.

G-P assume sages' view, that an inscription on a jug refers to the status of produce found in that jug. The inscription is probative, however, only if it explicitly names a consecrated status. If there is doubt as to the meaning of the inscription, the produce is assumed to be untithed. This is also sages' view in T. 5:1G-K.

4:12

A. One who says to his son, "[Coins in the status of] second tithe are in this corner,"
B. but he [i.e., the son] found [coins] in a different corner —
C. lo, these [i.e., the coins he finds] are deemed unconsecrated.
D. [If] there was there [i.e., in the corner specified by the farmer] a maneh [in the status of second tithe],
E. but he [later] found [there] two hundred [zuz] [i.e., two manehs] —
F. the extra [one maneh's worth of coins] is [deemed] unconsecrated.
G. [If there were there] two hundred [zuz in the status of second tithe],
H. but he found [there only one] maneh [i.e., one hundred zuz] —
I. it is all [deemed to be in the status of second] tithe.

M. 4:12 (y. Bes. 1:5; cf. b. Pes. 10a, b. Bes. 10b)

The pericope carries forward the concern of M. 4:9-11, determining whether or not, in cases of doubt, coins or produce are in the status of second tithe. In the three cases before us (A-C, D-F, G-I), someone has set aside a batch of coins in the status of second tithe. Subsequently, coins are found either in a different place (A-C) or of a different total value (D-F and G-I). We want to know whether or not these coins are considered to be the ones that were originally deposited, and are therefore to be deemed in a consecrated status. As we shall see, if there are grounds for assuming the coins are consecrated, we must treat them as such. This is the opinion of Judah in M. 4:10.

According to A-C, coins found in a location other than that in which consecrated money was left are not deemed to be consecrated. There is no reason for assuming that these are the same coins which were originally set aside. D-F and G-I, on the other hand, consider coins which are found where consecrated coins were left but which are of different total value from what was originally set aside. The point here is that an appropriate value from among these coins is deemed to be consecrated. In D-F the individual finds a greater value of coins than what was originally set aside. Since we have no grounds for claiming that the extra coins are in the status of second tithe, we deem consecrated from among these coins only a value equivalent to the value of the initial batch. At G-I the coins are of less value than the consecrated ones which had been set aside. We assume that all of these coins are part of the original batch, and therefore consider all of them to be in the status of second tithe.

A. One who says, "[The produce in the status of] second tithe is deconsecrated with the <u>sela^c</u> which will come up in my hand from this pocket,"

B. or, "with the <u>dinar</u> which will come up in my hand from this pocket,"

C. R. Yosé says, "He has not deconsecrated [the produce]."

D. And sages say, "He has deconsecrated [the produce]."

E. R. Yosé concedes to sages that in [the case of] one who says, "[The produce in the status of] second tithe is deconsecrated with the new <u>sela^c</u> which will come up in my hand from this pocket,"

F. or, "with the new <u>dinar</u> which will come up in my hand from this pocket,"

G. he deconsecrates [the produce].

H. Sages concede to R. Yosé that in [the case of] one who says, "[The produce in the status] of second tithe is deconsecrated with the <u>sela^c</u> which my son had (<u>šhyth byd bny</u>),"

I. or "with the <u>dinar</u> which my son had,"

J. he did not deconsecrate the produce.

K. For I say, "What if it were not in his [i.e., the son's] hand at the time [in which he spoke]?"

T. 4:12 (p. 266, ll. 44-50)
A-E: b. Er. 37b; H-K: y.
Dem. 7:5)

L. One who says, "[The produce in the status of] second tithe is deconsecrated with the sela^c which I will receive as change (š'prwṭ) for this dinar,"
M. or, "with the tressis which I will receive as change for this sela^c,"
N. makes change for (pwrṭ) the dinar and takes a sela^c from [the change],
O. [or makes change for] the sela^c and takes a tressis from the change],
P. [and the status of second tithe is transferred to the smaller coin].

T. 4:13a (p. 266, ll. 50-52)

In the dispute at A-D, a householder wants to transfer the status of second tithe from a batch of produce to a coin. Yosé and sages disagree as to whether or not the farmer must designate a particular coin to be consecrated in place of the produce. Yosé holds that the farmer must designate a specific coin while sages hold that this is not necessary.[32]

E-G and H-K expand upon the foregoing dispute. In E-G the householder does refer to a specific coin, one which he has in his pocket. Yosé therefore can agree that the farmer's declaration in this case has effect. H-K also have the farmer refer to a specific coin, but now in reference to its location. Since the coin may not actually have been where the householder claimed it was, his declaration may have referred to no coin at all. In this case, sages agree with Yosé that no transfer in status has occurred.

L-P returns to the formulary pattern of A-B. The householder now declares that he will transfer the status of second tithe to a coin which he will receive as change. The case here is similar to that of A-B in that no specific coin is being referred to. The validity of his declaration is in line with sages' opinion in D.

A. One who says, "I have [coins in the status of] second tithe in the house," but they are found in the attic;
B. [or he says, "I have coins in the status of second tithe] in the attic," but they are found in the house;
C. [or he says,] "They are in a bag," but they are found in a chest (dlysqys);

D. [or he says,] "They are in a chest," but they are found in a bag —

E. lo, these [coins are in the status of] second [tithe].

T. 5:4 (p. 267, ll. 14-16)

F. [one who says,] "I have [coins in the status of] second tithe," but he went and found them in the house or the attic or a bag or a chest —

G. lo, these [coins are] unconsecrated.

H. [One who says,] "I have in the house a bag [filled with coins in the status of] second tithe," but he went and found there two or three [bags full of coins] —

I. [the coins in] the largest are [deemed to be in the status of] second tithe and [the coins in] the two smaller [bags] are [deemed to be] unconsecrated.

J. He should not eat [produce purchased with coins] from the smaller [bags] until he deconsecrates them [i.e., transfers the status of second tithe in them to coins from] the large [bag].

K. [One who says,] "I have eight gold coins [in the status of second tithe] in a bag" and he went and found there (read with Lieberman: eight) gold <u>dinars</u> —

L. lo, these are [in the status of] second [tithe].

M. [One who says,] "I have eight gold <u>dinars</u> in the bag," but he went and found eight gold coins [other than <u>dinars</u>] —

N. lo, these are unconsecrated.

O. [One who says,] "I have eight gold coins in a bag," but he went and found fifty <u>sela^{c}s</u> or two hundred <u>zuz</u>,

P. two hundred <u>zuz</u> or fifty <u>sela^{c}s</u> [that is, coins equivalent in value to eight gold <u>dinars</u>],

Q. and afterwards found eight gold <u>dinars</u> —

R. lo, these [gold <u>dinars</u>] are [in the status of] second [tithe].

T. 5:5 (p. 269, ll. 16-23)

S. [One who says,] "I have eight gold <u>dinars</u> in a bag," but he went and found there fifty <u>sela^{c}s</u> or two hundred <u>zuz</u>,

T. two hundred <u>zuz</u> or fifty <u>sela^{c}s</u>

U. and afterwards found eight gold coins —

V. lo, these [gold coins] are [deemed to be] unconsecrated.

T. 5:6 (p. 269, l. 23 —
p. 270, l. 25)

W. [One who says,] "Lo, there is a maneh [in the status of second tithe]," but he went and found two hundred [zuz], that is two maneh's worth] [= M. 4:12D-E] —

X. "Unconsecrated [coins and coins in the status of] second tithe are mixed together," the words of Rabbi,

Y. but sages say, "All [of the coins] are unconsecrated."

Z. [If he says there are] two hundred [zuz], but he found [only] one maneh [= 4:12G-H] —

AA. "[We assume] one *maneh* was left behind and one *maneh* was carried off," the words of Rabbi.

BB. But sages say, "All [of the coins] are unconsecrated."

T. 5:7 (p. 270, ll. 25-27)

T. carries forward the concern of M. 4:12 with coins which a farmer claims are in the status of second tithe but which do not match his description. This essay can be subdivided into four units: A-G, H-J, K-V and W-BB, each of which contains a number of conflicting opinions.

In A-E a farmer declares that consecrated coins are in a specific location. In contrast to M., A-E declares that if the coins are found in a different location, they are deemed nonetheless to be in the status of second tithe. A-E claims therefore that any coins which might be the subject of the farmer's statement are deemed to be consecrated.[33] F takes the opposing view, in line with M. 4:12. According to this view, the coins must be in the exact location which the farmer mentioned in order for them to be deemed in the status of second tithe.

In H we find several batches of coins and do not know to which one the farmer referred. Two views follow concerning which batch is to be deemed consecrated. I claims that we assume that the largest batch is consecrated. According to J, however, we have no way of judging which batch of coins is consecrated. We therefore must go through the motions of transferring the status of second tithe from the smaller batches to the largest one. We choose the largest batch in order to consecrate the highest value of coins to which the farmer can be referring.

The third unit (K-V) considers cases in which the farmer claims that his consecrated coins are of one type, but another type is found. In contrast to A-E, T. now holds that the coins must match the farmer's description in order to be deemed consecrated. In K-L the farmer specifies only that he has gold coins. Gold dinars which are found where he claimed are clearly to be considered since these are gold coins. On the other hand, M-N has him specify that the consecrated coins are gold dinars. Other gold coins, which are not dinars and thus do not match his description, are not deemed to be consecrated. The same point is made at O-R and S-V, now with the coins in question mixed in with other coins. As in K-N, only the coins specifically described by the farmer are deemed to be consecrated.

Finally, W-BB cite M. 4:12. As we recall, M. ruled that if the farmer claimed the coins were of one value but a different value was found, the appropriate value of the found coins is deemed consecrated. Rabbi now provides a rationale for M.'s ruling (X-AA). In contrast, sages declare all of the coins to be unconsecrated. This is so because there is some doubt as to whether or not the farmer referred to these particular coins.[34]

A. [If] one was sick and could not talk (mšwtq),

B. they say to him, "Where are your coins in the status of second tithe,"

C. and he shows them with his fingers and thus and so (kk wkk)

D. [or if they ask him,] "Are [your consecrated coins] in such and such a place," [and] he nods with his head —

E. they question him three times.

F. If he answers consistently (^{c}l hn hn w^{c}l l'w l'w), his directions [dbryw] have effect (qyymyn).

T. 5:8 (p. 270, ll. 28-30)

G. [If] one was wondering about the location of ('ykn) [coins in the status of] second tithe [which belonged to his] father

H. [and] a man came to him in a dream and said, "They are thus and so," or "They are in such and such a place," —

I. it once occurred (zh hyh m^{c}śh) that they found coins there [where a vision predicted] and they went to ask the sages [about the status of the coins]. The sages said, "Lo, they are unconsecrated, for the words of a vision make no difference."

T. 5:9 (p. 270, ll. 30-33)
(y. M.S. 4:9, b. San. 30a)

T. continues its discussion concerning the categorization of coins which someone claims are in a consecrated status.

These two pericopae are formally related by their opening phrase hyh + participle and by their common theme. In each the description of the coins' location is accurate, but it is delivered in an unconventional way. In A-F the coins' owner cannot speak but directs someone to the coins by use of sign language. If his message is clearly understood, by being tested three times, the coins he refers to are deemed to be consecrated. G-I give us a case (G-H) followed by a precedent (I). Evidence supplied by a vision is not taken into consideration in matters of consecration since a vision has no recognized authority.

A. One who says to his son, "Even if you are dying [of hunger] do not touch [any coins in] this corner," and they went and found there coins —

B. lo, these are unconsecrated.

T. 5:10 (p. 270, ll. 33-34)

The farmer declares that certain coins are not to be used for ordinary purposes, implying that they are in a consecrated status. Nonetheless, the coins are deemed to be unconsecrated because the owner has not explicitly stated otherwise. The reasoning here is similar to that of sages in T. 5:1G-K.

A. [If] they [i.e., sons] saw their father hiding coins in a locker (śydh), a box or a chest,

B. and he said, "They are so and so's,"

C. [or he said], "They are [in the status of] second tithe" —

D. if he [makes his statement lightly] as if he was joking,

E. his statement has no effect (l' 'mr klwm);

F. if he [makes his statement with seriousness], as if he were intending to give testimony [in court],

G. his statement has effect (dbryw qyymyn).

T. 5:11 (p. 270, ll. 34-36)
(b. San. 30a)

H. [If] one said to them, "I saw your father hiding coins in a locker, a box or a chest [and the coins are in the status of second tithe],"

I. his statement has no effect [i.e., coins found in such a recepticle are deemed to be unconsecrated].

J. [If he says, "I saw your father hiding coins] in such and such a place, or in such and such a chest,"

K. his statement has effect [i.e., coins found there are deemed to be consecrated].

L. This is the general rule:

M. [Concerning] anything they can find [without his help] — (kl hmṣwy lhm) — his statement has no effect.

N. [Concerning] anything they cannot find [without his help] — his statement has effect.

T. 5:12 (p. 270, l. 36 — 271, l. 39)
(b. San. 30a)

O. [If] a launderer says to him, "This cloak was your father's, but I bought it from him,"

P. he is believed (h'mn),

Q. for he who has the power to prohibit has the power to permit (hph š'sr hw' hph šhtyr).

R. But if there were witnesses to the fact that it was his father's and he said, "I bought it from him,"

S. he is not believed.

T. 5:13a (p. 271, ll. 39-41)
(O-S: cf. M. Ket. 2:2)

On formal grounds we have four units: A-G, H-K, L-N and O-S. A-G and H-K are linked at A and H and through the use of the standard formulas "his statement has effect/no effect" at E, G and I, K. Although H-K and L-N do not explicitly mention second tithe, the context of this unit requires that this be the issue. The general rule at L-N is formally disjunctive with H-K, shifting the subject from the witness to the sons.[35] O-S is formally autonomous of the foregoing but makes the same point.

The issue again is the validity of a statement for deciding whether or not an item is deemed to be in the status of second tithe. T. takes up an issue not considered in M., the trustworthiness of the speaker. A-G declare, first of all, that if there is any reason to doubt the sincerity of the speaker, his statement is disregarded. If part of a witness' testimony works to his own detriment, however, we assume the entire statement is reliable. In H-N, for example, someone describes the location of certain coins and claims that they are consecrated. If his description is detailed, and so allows us to find the coins, we accept as well his claim that the coins are consecrated. The assumption is that if he wished to prevent the use of the coins, he would not have disclosed their location.[36] In O the launderer himself tells us that the cloak once belonged to someone else. Since he himself volunteered the information that the cloak was not originally his own, we can believe his claim that he bought it from the original owner.

CHAPTER FIVE
MAASER SHENI CHAPTER FIVE

Having completed the discussion of produce in the status of second tithe, the tractate moves to a consideration of two categories of food which are in a status similar to that of second tithe. The first of these is fruit of a tree's fourth year of growth (M. 5:1-5). Scripture (Lev. 19:23-25) assumes that such fruit is both holy and may be eaten by the farmer, a fact which permits M. to draw a pertinent analogy to second tithe (M. 5:3). The second category of produce is agricultural gifts of which the farmer has not properly disposed (M. 5:6-15). Such consecrated produce, which accumulates uneaten in the farmer's house, is like second tithe which, in the era after the destruction of the Temple, cannot be brought to Jerusalem and, therefore, also remains uneaten in the farmer's possession. With this larger structure of the chapter in mind, let us consider each of its parts in greater detail.

The basic facts concerning produce of a tree's fourth year of growth (M. 5:1-5) come from Scripture. Lev. 19:23-25 declare that fruit growing on a tree "shall be forbidden to you (yhyh lkm ᶜrlym) for three years and is not to be eaten, but in the fourth year its fruit is holy, dedicated to the Lord." Scripture makes it clear in this passage, first of all, that produce of a planting's fourth year is consecrated. Furthermore, Scripture draws a contrast between produce of the tree's first three years, which is not eaten, and produce of the fourth year, for which this claim is not made. From this contrast, the Hillelites (M. 5:3) conclude that while fruit of the first three years is not eaten, that of the fourth is eaten. This being the case, they see an analogy between fruit of a tree's fourth year of growth and produce in the status of second tithe. Both of these are consecrated foods eaten by the farmer. On the basis of this analogy, they rule that in other respects produce of the fourth year is to be treated as is produce in the status of second tithe. If it cannot be eaten in Jerusalem, for example, it must be destroyed. Furthermore, the farmer who transfers the status of second tithe from his own produce to his own coins, in order to bring the coins to Jerusalem, must pay the added fifth. Finally, it may not be left for the poor, since it must be eaten by the farmer himself. In each of these cases, the Shammaites disagree. They hold that since Scripture does not specify that produce of the fourth year is eaten, it is not analogous to second tithe at all. They rule, therefore, that fruit of the fourth year is never treated like second tithe.

The Hillelite notion that there is an analogy between fruit of a tree's fourth year of growth and produce in the status of second tithe is assumed by the other pericopae in this unit. M. 5:1-2 declare that fruit of the fourth year is used in the same way as is second tithe. That is, it is brought to Jerusalem for consumption or must be sold and money brought to the city to purchase other produce in its stead. M. 5:4-5 assume that fruit of the fourth year may be sold and go on to discuss the next logical question — how its selling price is established. In particular, what interests M. is how the price is set in unusual cases, such as when the produce is unharvested (M. 5:4) or when the produce has no owner (M. 5:5). This material is parallel to M. 4:1-8's discussion of how one establishes the selling price of second tithe.

The second unit of the chapter (M. 5:6-15) turns our attention to agricultural gifts of which the farmer has not yet properly disposed. According to M. 5:6, the produce either must be given to its proper recipient or must be destroyed. The notion emerges from Deut. 26:12-15, which states, "When you finish separating all the tithes from your produce [in the] third year, the year of tithing, you shall give it to the Levite, the stranger and the orphan and the widow . . . then you shall declare before the Lord your God, 'I have removed [all] consecrated produce from my house . . .'" The problem for M.'s formulators is that Scripture here states that tithes are separated only every three years. This contradicts M.'s assertion that the farmer separates agricultural gifts annually. In order to overcome this contradiction, the formulators of the law read Scripture to mean that every three years the farmer must give to their appropriate recipients only those agricultural gifts which he previously separated but which still remain in his house. At this time, all heave-offering must go to the priests, first tithe must be given to the Levites, second tithe must be eaten in Jerusalem and poorman's tithes must be given to the poor. M. 5:6 rules that with the Temple's destruction, when second tithe no longer could be eaten, such produce is removed from the farmer's domain by being destroyed.

With the basic law of removal set forth in M. 5:6, the redactor turns to a consideration of two related themes: how the removal is carried out in special cases (M. 5:7-9) and the meaning of the confession the farmer recites in Jerusalem (Deut. 26:13ff, M. 5:10-14). In this later block of material, the text of the confession is the subject of a midrashic interpretation which stresses the importance of properly carrying out the tithing laws.

The subunit closes with an appendix which refers to a series of legal enactments ascribed to the time of Yoḥanan, the High Priest (M. 5:15). This material is in the subunit because the first of these acts refers to the confession. M.'s discussion has already drawn to a natural close, however, at M. 5:14.

5:1

A. (1) [As regards] a vineyard in its fourth year [of growth]—
B. they mark it off with clods of earth.
C. (2) And [a vineyard] in its first three years of growth (^{c}rlh) [they mark it off] with clay.
D. (3) And [an area] of graves [they mark off] with lime
E. which they dissolve in water and pour out [along the boundary] (wmmḥh wšwpk).
F. Said Rabban Simeon b. Gamaliel, "To what [case] does this apply?
G. "During the Sabbatical year."
H. "And those who are conscientious (hṣnw^{c}ym) set aside coins and say, 'Let whatever is plucked [from this vineyard which is in its fourth year] be deconsecrated with these coins.'"

M. 5:1
(b. B.Q. 69a; H: b. B.Q. 68b)

The opening triplet provides a list of three areas which must be marked off as a precaution on behalf of unwitting passers-by.[1] Only the first two rules are concerned with food. A-B assumes that fruit produced by a vineyard in its fourth year of growth is consecrated and may be eaten only in Jerusalem (Lev. 19:23-25). The ruling therefore requires landowners to mark such fields so that strangers will not pick the produce and eat it as though it were unconsecrated. Grapes of a vine's first three years of growth, the subject of C, are altogether forbidden for use as food.[2] As at A-B, the owner of the vineyard must warn passers-by of the status of the produce, lest they pick the grapes and eat them in violation of the law. D-E shift our attention to graveyards. These render unclean any priest who enters their bounds.[3] These areas must be marked off, therefore, so that priests will not unwittingly cross into such areas and be contaminated.

Simeon b. Gamaliel's opinion (F-G) refers to the produce of A-C. His point, according to Maim., Bert., and TYY, is that normally we do not need to worry that passers-by will eat grapes from a vineyard without the owner's permission. This becomes a consideration only during the Sabbatical year, when the produce is legally ownerless. Now strangers may indeed feel free to help themselves to the growing fruit. It is during this year, then, that the farmer must mark off the grapes as being in a consecrated status. I presume that his rule is placed here, rather than after C, so as not to break up the triplet at A-C.

H offers a second way of insuring that consecrated grapes are not picked and wrongfully eaten. The farmer sets aside coins for the express purpose of taking on the status of any consecrated grapes that are picked.[4] Grapes which a passerby picks therefore are immediately deconsecrated and may be eaten as ordinary produce.[5] The coins then must be used by the farmer in Jerusalem to purchase other produce to be eaten in place of the original grapes. Through this ruling, the pericope introduces, in an oblique manner, the topic of Chapter Five, produce grown during a vine's fourth year.

T. Others say,

(1) "[A tree bearing produce] dedicated [to the Temple] they mark with red dye.

(2) "A place of idolatrous worship they blacken with charcoal.

(3) "On a leprous house they throw wood ash.

(4) "A place of execution they mark off with blood.

(5) "The place in which a calf's neck is broken [to make atonement for an unsolved murder] they mark out with a circle of stones."

T. 5:13b (p. 271, ll. 41-44)
(y. M.S. 5:1)

The pericope supplements M. 5:1A-D's list of places which must be marked off to prevent individuals from using them improperly. In each case, the material used to designate the prohibited item is symbolic of the nature of the prohibition. Items dedicated to the Temple (1) may not be eaten by commoners, but must be handed over to Temple authorities. They are marked off with red dye, a reminder of the red line which divided the Temple altar into upper and lower portions (see M. Mid. 1:3).[6] Pagan shrines (2) are not to be entered by Israelites (M. A.Z. 3:7). They are marked off with ash to indicate that they are to be burnt to the ground (cf. T. A.Z. 4:19). A house which is contaminated with leprosy-like spots (3) renders unclean whoever enters it (M. Neg. 13:4). Since such a house is to be reduced to dust (Lev. 14:45, see M. Neg. 12:7), it is covered with dust. A place of execution is marked off to prevent passers-by from coming into contact with a dead body and so being rendered unclean. The blood used to signify this area symbolizes the nature of the uncleanness adhering to the site.[7] Atonement for an unsolved murder is made by breaking the neck of a young calf (Deut. 21:1-9). The site of the ceremony is set off by stones to indicate that the site may not be used for agriculture (Deut. 21:4).[8]

The phrase "others say," which introduces the pericope is meaningless here, since the material it introduces is substantively independent of what proceeds.

5:2

A. [Produce of] a vineyard in its fourth year [of growth] was brought (hyh ᶜwlh) to Jerusalem [if it was grown] within one day's [journey of the city] in any direction [i.e., such produce could not be sold and the coins brought in its stead].

B. And what is the extent [of a day's journey from Jerusalem]?

C. Elat to the south,[9] 'Aqrabah to the north, Lod to the west and the Jordan [River] to the east.

D. But when produce became [too] abundant, they ordained (htqynw) that [the produce] should be redeemed [i.e., sold] [even if it grew] a short distance from the city wall (smwk lḥwmh) [of Jerusalem, and the farmer would bring coins into the city instead of produce].

E. But there was a stipulation (wtny hyh hdbr) that whenever they wanted [to reverse their decision] the law (dbr) would revert to its original form (lkmwt šhyh).

F. R. Yosé says, "When the Temple was destroyed, this stipulation (D) was made.

G. "And there was a [further] stipulation, that whenever the Sanctuary would be rebuilt, the law would revert to its original form."

M. 5:2
(A-C: b. R.H. 31b;
B. Beṣ. 5a)

The pericope is built around two opposing views, A and D+E. B-C gloss A, and F-G offer a second rationale for D's law. The point of the pericope emerges from the contrast between A and D. Under discussion is whether or not the farmer may sell grapes of a vineyard's fourth year of growth so that he can bring money to Jerusalem in place of the fruit. According to A, the farmer may sell such grapes only if they are grown further than one day's journey from the holy city. The reason, presumably, is the fear that the grapes will not survive a long trip to Jerusalem but will spoil. Since the farmer cannot bring these grapes to Jerusalem, A allows him to sell them and bring coins in their stead. Grapes grown closer to the city, however, can be brought there without spoiling and so they may not be sold. D disagrees with A's restriction, declaring that all grapes of a vine's fourth year may be sold, irregardless of where they are grown. D-E give an economic explanation for this law. According to D, too much produce was being brought into the city, thereby flooding the market. The effect of the law at D is to allow the farmer to sell the fruit outside of Jerusalem, thereby cutting down the amount of produce brought into the city. Yosé, F-G, offers another explanation for D's law. According to him, the law was changed in response to the Temple's destruction. Now no consecrated produce could be eaten in the city. The law therefore was changed so that the farmer could sell his consecrated grapes no matter where they were grown.

A. [Produce of] a vineyard in its fourth year [of growth] is brought to Jerusalem [if grown] within one day's journey [of the city in any direction] [M. 5:2A].

B. To what [case] does this apply?

C. [To a case in which] the vineyard has [at least] five vines.

D. But [as regards] a vineyard that does not have [at least] five vines,

E. and [likewise as regards groves] of other [types of] trees,

F. lo, this [produce, when grown during the fourth year of growth] may be redeemed even if grown a short distance from the wall [of Jerusalem] [cf. M. 5:2D].

G. Rabban Simeon b. Gamaliel says, "It is the same as regards a vineyard which has five vines and as regards a vineyard which does not have five vines,

H. "lo, this [produce] must be brought to Jerusalem [and may not be sold]."

I. [And when such fruit is brought to Jerusalem], he divides it among his neighbors, his relatives and his acquaintances and he decorates the public thoroughfare (šwq) with it.

J. Said R. Simeon, "There is no decorating the public thoroughfare [with such produce].

K. "Rather, he redeems [the fruit] at the prevailing market value in his locale and [produce purchased with] its proceeds are eaten [in Jerusalem just as in the case of] produce in the status of second tithe (reading with E: km^{c}śr sny)."

T. 5:14 (p. 271, ll. 44-49)

According to M.'s rule, grapes grown in a vineyard's fourth year may be sold only if they are grown further than a day's journey from the city. B-F limits the scope of this law by claiming that it applies only to grapes grown in a true vineyard, that is, a plot of land in which at least five vines are growing (cf. M. Kil 4:5). In all other cases, the grapes may be sold irregardless of where they grow. By liberalizing the law in this way, T. achieves much of the same results as M. 5:2 D-G, but without claiming the law was changed. Simeon b. Gamaliel resists this attempt to weaken A and maintains that the law applies to all grapes, and presumably to other fourth year produce as well.

I is autonomous of the foregoing. It claims that one of the permissible uses of such grapes is to beautify the city.[10] Simeon (J-K) disagrees with both A-H and with I. His reasoning, stated explicitly in K, is that grapes grown in a vine's fourth year are to be treated like produce designated as second tithe. The farmer always has the right, therefore, to sell such grapes and bring coins to Jerusalem in their stead (F). Furthermore, the farmer must eat in Jerusalem produce of a vine's fourth year, just as he must eat in the city produce in the status of second tithe. Such produce is not to be used for other purposes (J).

A. When the Temple was destroyed, the first court (byt dyn hr'-šwn) said nothing about it [i.e., about changing the law as regards bringing to Jerusalem fruit grown during the fourth year of a vineyard's growth].

B. The later court decreed that this [fruit] is redeemed even if grown close to the city wall [of Jerusalem] [M. 5:2D].

T. 5:15 (p. 271, 11. 49-51)

C. M^{c}śh b R. Eliezer owned a vineyard on the border of Kefar Tabri to the east of Lod and did not want to redeem [the fruit grown during the fourth year of the vines' yield, but wanted to keep the produce itself until the time of removal when he would have to destroy it].

D. His disciples said to him, "Rabbi, since [the court] decreed that this [fruit] is redeemed even if grown close to the city wall [of Jerusalem], you must redeem it."

E. R. Eliezer immediately harvested [the grapes] (^{c}md wbṣr) and redeemed them.

F. And [as regards the fruit of] all other trees, [the law that governs] what grows during the tree's fourth year [of maturity] is like [the law] of second tithe.

T. 5:16 (p. 271, 1. 51 — 272, 1. 54)

A-B restate the law of M. 5:2D-G, but without taking a position on why the law was enacted. Although A claims that the change occurred after the Temple's destruction, as Yosé did in M. 5:2F-G, the change is not specifically linked to this fact. C-E exemplify the law's effect. Eliezer wanted to keep in his possession grapes grown in the fourth year of the vineyard, but is told that he must sell the grapes instead, transferring their value to coin.

F is an autonomous unit of material. It declares that the law which applies to grapes of a vineyard's fourth year of growth applies also to produce grown during the fourth year of other fruit trees as well.

5:3

A. [As regards] a vineyard in its fourth year of growth —
B. the House of Shammai say, "[The law of] the added fifth does not apply (<u>'yn lw ḥmš</u>), and [the law of] removal does not apply."
C. And the House of Hillel say, "[The laws of the added fifth and of removal] do apply (<u>yš lw</u>)."
D. The House of Shammai say, "[The laws requiring leaving] single grapes [for the poor] apply, and [the law requiring leaving] defective grape clusters [for the poor] apply.
E. "And the poor [who gather such grapes] redeem them themselves [by transferring the consecrated status of the grapes to coin]."
F. And the House of Hillel say, "[The owner of the vineyard must himself bring] all [of the grapes grown in the fourth year] to the winepress [i.e., the laws of single and defective grapes do not apply]."

M. 5:3
(M. Pe. 7:6; M. Ed. 4:5;
A-D+F: b. Ket. 54b; D-F:
Sifra Qod. par. 3:7)

The two formally autonomous Houses' disputes (B-C and D+E-F) depend on a single superscription (A). At issue in both disputes is whether or not the laws distinctive to produce in the status of second tithe apply as well to the crop of a vine's fourth year of growth. This question arises because in both cases the consecrated produce must be eaten in Jerusalem by the farmer. As we shall see, the Hillelites claim that whatever is true of second tithe is therefore also true of fourth year produce. The Shammaites hold that the two types of consecration are distinct. They therefore argue that no analogy can be drawn from the one to the other.

We turn first to the laws of the added fifth and of removal (A-C). The law of the added fifth requires a farmer who transfers the status of second tithe from his own produce to his own coins to pay an extra charge of twenty-five percent (M. 4:3). The law of removal comes into effect during the fourth and seventh years of the seven-year Sabbatical cycle. Before Passover of these years, the farmer must remove all consecrated produce from his domain (M. 5:6). The Hillelites claim that just as these laws apply to produce designated as second tithe, so too do they apply to grapes of a vineyard's fourth year (C). The Shammaites, as we have said, hold that the status of such grapes is distinct from the status of second tithe. It follows, in their view, that the laws of the added fifth and of removal do not apply to such grapes (B).

The subject of D-F is Scripture's requirement to leave aside a portion of one's crop for the poor (Lev. 19:10). These laws do not apply to second tithe since the farmer must himself bring the entire tithe to Jerusalem. As before, the Houses disagree as to whether or not these laws are applicable to produce of a vine's fourth year. The Shammaites again argue that what is true of second tithe is not true of produce of the fourth year. It follows that the laws of single and defective grapes do apply to grapes of a vine's fourth year. E makes the obvious point that since the grapes are nonetheless consecrated, the poor who receive them must deconsecrate them or eat them in Jerusalem.

The Hillelite opinion in F is problematic. We would expect this lemma to read, "[The laws of . . .] do not apply." Instead, F introduces a new notion, that the farmer must process his fourth year produce into wine before bringing it to Jerusalem. I cannot explain why this view has been introduced here. The point of the lemma for the dispute, however, is clear. The farmer processes all of his consecrated grapes for use in Jerusalem. He may not leave any aside for the poor. Single grapes or defective grapes are not to be left for the poor, just as is the case for grapes declared to be second tithe.[11]

A. [As regards] a vineyard in its fourth year of growth —

B. the House of Shammai say, "[The law of] the added fifth does not apply and [the law of] removal does not apply."

C. And the House of Hillel say, "[The law of] the added fifth applies and [the law of] removal applies [M. 5:3A-C]."

D. "Under what conditions do these [rules,i.e., the Shammaite opinion] apply?

E. "During the Sabbatical year.

F. "But during the other years of the Sabbatical cycle, [the law of] the added fifth applies and [the law of] removal applies,"

G. the words of Rabbi.

H. Rabban Simeon b. Gamaliel says, "[As regards] both the Sabbatical year and the other years of the Sabbatical cycle —

I. "The House of Shammai say, '[The law of] the added fifth does not apply and [the law of] removal does not apply.'

J. "And the House of Hillel say, '[The law of] the added fifth applies and [the law of] removal applies [M. 5:3A-C].'"

T. 5:17 (p. 272, ll. 54-59)
(A-C: M. Pe. 7:6; M. Ed. 4:5;
b. Ket. 54b; D-G: y. M.S. 5:3;
E-I: y. Pe. 7:6)

Rabbi, D-F, brings the Shammaites into essential agreement with the Hillelites by claiming that the Shammaite view applies only during the Sabbatical year. During the Sabbatical year, all produce which grows is deemed to be ownerless. Rabbi's point is that since the farmer is not the owner of produce during this particular year, he does not pay the added fifth when transferring the status of the produce to his own coins, and he is not responsible for removing the produce from his domain before Passover. Since in the other years of the Sabbatical cycle the farmer is considered to be the owner of his crop, Rabbi claims that the Shammaites agree that he is subject to these laws, just as the Hillelites hold. Simeon b. Gamaliel, on the other hand, argues that the Shammaite opinion is not qualified in this way and that the Houses do indeed stand in opposition to each other.[12]

5:4-5

A. How do they redeem [i.e., sell] [produce from a] planting's fourth year of growth (nṭ^{c} rb^{c}y)?

B. (1) He sets the basket [of produce] in front of three [potential buyers] and says, "How many [baskets of such produce as yet unharvested and still in the field] is one willing to redeem for a sela^{c}, on condition that (^{c}l mnt) he [that is, the purchaser] pay the expenses [of harvesting the produce] out of his own pocket (mbytw)?"

C. Then (w-) [the one who purchases the produce] sets aside money [with which to purchase the produce] and says, "All produce of this [type, i.e., the type in the basket] which is picked [at my expense] is [to be deemed] deconsecrated with these coins at [a rate of] so many baskets to a sela^{c} ((bkk wkk slym bsl^{c})."

M. 5:4
(cf. M. San. 1:3)

D. (2) But during the Sabbatical year [the farmer] redeems (i.e., deconsecrates produce of the fourth year] at its full value [that is, the purchaser may not deduct from the selling price what it will cost to harvest it].

E. (3) And if [during the other years of the Sabbatical cycle] there is an entire [crop which has been declared] ownerless, [and someone wants to sell it and bring the consecrated coins to Jerusalem, the one who deconsecrates it receives] compensation only for [what it costs him to] harvest [the crop, that is, he does not receive the full market value of the produce].

F. One who redeems his own produce from a planting's fourth year [so as to deconsecrate it] adds a fifth [to its selling price],

G. whether [the produce] was his [originally] or was given to him as a gift [cf. M. 4:3C-D].

M. 5:5 (b. Qid. 54b; y. M.S. 4:3; cf. M. B.M. 4:8)

Fruit grown during the fourth year of a tree's growth, like grapes of a vine's fourth year, must either be eaten in Jerusalem or be sold, with the proceeds brought to the holy city.[13] Since the legal status of this produce is like that of grapes of the fourth year, M.'s consideration of this topic simply carries forward its previous discussion. Of specific concern here is how the price for the deconsecration of such produce is determined. We have already seen that the price for redeeming consecrated produce is the same as the selling price of unconsecrated produce of its same type (M. 4:1-3). The discussion here considers how this procedure for setting prices works in certain special cases in which the procedure used for second tithe offers no precedents. We consider first, in B-D, produce which is sold before it is harvested, either because the farmer chooses to do so (B-C) or because the laws of the Sabbatical year prohibit such labor (D). E considers consecrated produce which has been declared ownerless. F-G, finally, deal with produce which a farmer wants to exchange for his own money.

In B-C the farmer decides to sell his produce while it is yet unharvested. By so doing, he saves himself the effort of investing further labor in a crop which he cannot fully enjoy, since he must eat the produce, or its equivalent value in other food, in Jerusalem. The point here is that since the unharvested produce is worth less to the buyer than is harvested produce, a smaller amount of money is consecrated in its place. The farmer benefits from selling the produce still in the field because he has less consecrated money to spend in Jerusalem than if he sold the crop after its harvesting.[14]

D has the fourth year of the plant's growth coincide with the Sabbatical year. The problem here is that during this year, all agricultural labor, such as harvesting, is prohibited.[15] All produce grown during this year, therefore, must be sold in an unharvested state. Since, in this regard, the unharvested consecrated produce is no different from other produce on the market, it is sold at the going market rate.

E offers a third complication. Now the produce has been declared ownerless so that no one is responsible for properly disposing of it. Someone, however, takes upon himself the responsibility of harvesting the crop, selling it and spending the coins in Jerusalem. E declares that this person may sell the produce only for the cost of harvesting.[16] Since the crop is ownerless and could have been taken by anyone, it has no market value in itself. Only by harvesting the crop has the seller added value to it and, it follows, he may sell it only for the value of the labor he invested in harvesting it. The farmer is required to spend only this lesser amount of money in Jerusalem.

F-G concludes M.'s discussion by applying to fourth year produce the same rule M. 4:2 applies to produce in the status of second tithe, namely that a farmer who is transferring the consecrated status of the produce to his own coins pays the added fifth. The rule as stated here is copied verbatim from M. 4:3C-D, except for the substitutes of "second tithe" in place of "produce from a planting's fourth year."

I. A. The House of Shammai say, "They do not prune it [i.e., a vineyard in its fourth year of growth]."

B. And the House of Hillel say, "They do prune it."

T. 5:18 (p. 272, l. 59)

II. C. The House of Shammai say, "They do not redeem [i.e., deconsecrate by selling produce from a vineyard's fourth year of growth when it is still in the form of] grapes but only [after it is processed into] wine."

D. And the House of Hillel say, "[They redeem such produce either as] wine or as (w-) grapes."

E. However ('bl), both [Houses] agree that they do not redeem [i.e., deconsecrate produce from a vineyard's fourth year of growth while it is still] attached to the ground.

T. 5:19 (p. 272, ll. 59-61)

III. F. The House of Shammai say, "They do not plant [a tree or vine] during the fourth year [of the Sabbatical cycle],

G. because produce of the fourth year [of its growth] will coincide (ḥl lhywt) with the Sabbatical year."

H. But the House of Hillel permit [planting trees or vines during the fourth year of the Sabbatical cycle].

T. 5:20 (p. 272, ll. 60-61)

These disputes deal with a single theme: whether or not the farmer is obligated to increase the overall value of his consecrated fourth-year crop. In each case, as we shall see, the Shammaites hold that a farmer must insure that the crop will fetch its highest possible selling price. The Hillelites, on the other hand, hold that the farmer is under no such obligation. The pericopae are formally unitary, constituting a triplet, based on the parallel opinions of the Shammaites, A, C, and F.

A-B consider whether or not the farmer may prune back his vines at the beginning of their fourth year of growth. This would increase their yield in future years.[17] In so doing, however, he destroys vines which otherwise would produce a consecrated crop during that growing season. The Shammaites claim that the farmer therefore may not trim the vines. They must be allowed to realize their fullest yield. The Hillelites, on the other hand, hold that since consecrated grapes have not yet appeared on the vine, the farmer has no obligation to preserve the plant. He thus may prune back the vines.[18]

C-E consider the point in their growth and processing at which grapes of a vineyard's fourth year may be sold. The Shammaites rule that the produce is sold only after it has been processed into wine. The wine fetches a much higher price on the market than do raw grapes. The point, as at A, is that the farmer should receive for his consecrated crop its highest possible market price.[19] The Hillelites allow the farmer to sell even unprocessed grapes. This is in line with their view at B that the farmer is not required to get the highest possible price for his consecrated grapes. C has the Houses agree that the consecrated grapes are sold only after they are harvested. The law here regards grapes of a vine's fourth year to be analogous to pieces of fruit designated as second tithe, which also are sold only after they are harvested (M. 1:5). The notion that the status of fourth year growth is like the status of second tithe has been attributed to the Hillelites in M. 5:3. T.'s rule clearly contradicts what is assumed in M. 5:4-5.[20]

F-G concludes the triplet by asking whether or not the farmer should avoid planting in the fourth year of the Sabbatical cycle so as to keep consecrated fourth year produce from growing during the Sabbatical year. The point is that such produce may not be cared for or processed during the Sabbatical year, since all agricultural labor is prohibited at this time. The Shammaites will not want fourth year produce to grow this year, since they want such grapes to be processed into wine before being sold, and this cannot now be done. They therefore rule that the farmer must avoid having consecrated produce growing during this year.[21] The Hillelites again do not obligate the farmer to produce a valuable crop of consecrated produce. Their position is clearly presupposed by M. 5:5D.

5:6

A. The day preceeding (^{c}rb) the (read with most mss: 'ḥrwn)[22] last festival-day of Passover in the fourth and seventh [years of the Sabbatical cycle] was [the time of] removal [i.e., the time at which the farmer must remove all consecrated produce from his domain].

B. How was [this] removal [carried out]?

C. (1) They give [produce in the status of] heave-offering and [in the status of] heave-offering of the tithe to those entitled to it (<u>lb^clym</u>) [i.e., the priests],

D. (2) and [they give produce in the status of] first tithe to those entitled to it [i.e., the Levites],

E. (3) and [they give produce in the status of] poorman's tithe to those entitled to it [i.e., the poor].

F. (4) But [produce in the status of] second tithe and first fruits are removed [i.e., destroyed] (<u>mtb^crym</u>) under all circumstances [i.e., even outside of Jerusalem, since the farmer has no time to take them to the city].

G. R. Simeon says, "First fruits are given to the priests [even outside of Jerusalem], just as [in the case of produce in the status of] heave-offering."

H. A cooked dish [containing produce in a consecrated status] —

I. the House of Shammai say, "He must remove it [at the time of removal]."

J. And the House of Hillel say, "Lo, it is as if it were already removed [since it has been transformed by the cooking]."

M. 5:6
(Cf. Sif. D. 109; 302)

The pericope inaugurates a block of materials (M. 5:6-9) which spells out when and how the farmer carries out the law of removal. This law requires the farmer periodically to remove from his domain all agricultural gifts which he has designated but not yet disposed of and which therefore have gathered in his house. At the time of removal, all such consecrated produce must either be given to its proper recipients: priests, Levites or the poor as appropriate, or it must be destroyed. In this way, the farmer is prevented from indefinitely holding back offerings which belong to others. This law is based on Deut. 26:12f, which declares, "When you finish separating all the tithes from your produce in the third year, the year of tithing, you shall give it to the Levite, the stranger and the orphan and the widow . . . . Then you shall declare before the Lord your God, 'I have removed [all] consecrated produce from my house . . . .'" On the basis of these facts, A-F+G describe the procedure for fulfilling this obligation. The Houses' dispute in H-J takes up a secondary issue: whether or not consecrated foods which have been cooked are subject to this law of removal.

We turn first to M.'s declaration that the removal must be completed by Passover of the fourth and seventh years of the Sabbatical cycle (A). These times certainly are derived from the Scriptural statement that removal is to be effected after the harvest of every third year. M. understands Scripture to mean that removal is carried out after all produce planted during the third and sixth years of the Sabbatical cycle is harvested. This requires that the time of removal occur during the fourth and seventh years of the Sabbatical cycle, after the fall harvest.[23] Passover of these years is the logical time for the removal since it is the first pilgrimage festival after this harvest. M. assumes that the farmer will be in Jerusalem for this festival and therefore requires that at that time he recite the required confession, stating that he has properly carried out the removal.[24]

B-G turn our attention to what is to be done with the various gifts when the time of removal arrives. Agricultural gifts which are designated for particular recipients, priests or Levites, for example, must be given to those persons.[25] Produce which must be eaten in Jerusalem by the farmer, such as produce in the status of second tithe, is destroyed (F). This is so because the farmer is no longer allowed to have such food in his possession and so may not make use of it in the city. G disputes F's notion of the proper disposal of first fruits. These are the first pieces of fruit which are harvested each year. They are designated by the farmer as consecrated offerings and are taken to Jerusalem where they must be eaten by priests (M. B.K. 2:1). This fruit, then, is both like heave-offering, which is given to the priests, and like produce designated second tithe, which is eaten in Jerusalem. There are, consequently, two analogies available for determining the proper distribution of this produce at the time of removal. On the analogy of second tithe, it would seem that first fruits which are not eaten by the time of removal must be destroyed. This is the view of F.[26] Simeon (G) holds, on the other hand, that first fruits are to be treated like heave-offering, since they are the property of the priests. At the time of removal they must be given to the priests, he holds, just as is the case for food in the status of heave-offering.

At H-J the question is whether or not consecrated produce which has been cooked is subject to the law of removal. The question arises because the food is transformed in cooking.

The Shammaites hold that since the consecrated ingredients are nonetheless present in the dish, the law of removal applies. The Hillelites disagree, and their reasoning is given in J. They hold that the food is so altered in the process of being cooked that we deem it already destroyed and so no longer subject to the law of removal.[27] I cannot explain why this should be so, however, since M. nowhere else deems cooking to destroy consecrated food.[28]

A. A tree which they planted on the day preceeding the New Year's day of the Sabbatical year —

B. [what is separated from the tree as] second tithe is not [subject to the law of] removal until the Sabbatical year [which follows].

T. 5:21 (p. 272, ll. 62-63)

Second tithe is not separated from the fruit of a tree until the tree's fifth year of growth. This is so because fruit grown prior to this time is either in the status of <u>ᶜorlah</u> (first three years) or is consecrated as produce of the fourth year (M. 5:1-5). The tree described here has been planted at such a time that agricultural gifts are first taken from the tree only during the fifth year of the Sabbatical cycle, after the time of removal of the fourth year has passed. The first time these pieces of fruit are subject to the law of removal, consequently, is Passover of the coming Sabbatical year (M. 5:6).

A tree normally enters a new legal year of growth each fifteenth of <u>Shevat</u>, the new year of trees (M. R.H. 1:1). An exception is made, however, if the tree is planted less than six months before this date. A tree less than half a year old on the fifteenth of <u>Shevat</u> is not deemed to enter a new year of growth until the following fifteenth of <u>Shevat</u>, twelve to eighteen months after its planting. This is the case before us. The tree now has been planted just before the beginning of the Sabbatical year, only four and a half months before the fifteenth of <u>Shevat</u>. As a result, the tree is deemed to enter its second year of growth on the fifteenth of <u>Shevat</u> of the first year of the new Sabbatical cycle. Its fifth year of growth, consequently, begins only in the middle of the fourth Sabbatical year. Fruit appearing on the tree subsequent to this time will be ready for picking only after the time of removal in the fourth year has passed.[29]

5:7

A. One who had produce [in the status of second tithe] at this time [i.e., after the destruction of the Temple][30]

B. and the time for removal arrives —

C. the House of Shammai say, "He must deconsecrate [the produce by transferring its consecrated status] to coin."

D. And the House of Hillel say, "It is all the same (<u>'ḥd...'ḥd</u>) whether [the farmer removes his consecrated food in the form of] coin or [in the form of] produce."

M. 5:7

Scripture declares, "If you are unable to carry it [i.e., produce in the status of second tithe, to Jerusalem] because the place [i.e., Jerusalem] where the Lord your God has chosen to cause his name to dwell is too far from you, sell the produce for coins . . . (Deut. 14:24-25)." The Houses now consider the implications of this verse for the situation after 70.

We recall that after the Temple's destruction, the farmer cannot bring consecrated produce to Jerusalem for consumption (M. 1:5-6). At issue is whether or not on that account we demand that the farmer sell his consecrated food, as Scripture would suggest. The Shammaites claim that since it is impossible to bring the produce to Jerusalem, we must honor Scripture's demand that the produce be sold. This sale takes place at the time of removal because at this time the farmer must get rid of all consecrated food in his domain (M. 5:6). The Shammaites, however, do not describe what is to be done with the consecrated coins, a problem which explains the Hillelite position at D. In the Hillelite view, the point of Scripture is to allow the farmer to bring to Jerusalem money to be used to purchase food which will be eaten as second tithe. With the city destroyed, however, such consecrated coins themselves cannot be properly spent. In all events, then, the farmer is left with consecrated items which he cannot use. Since the sale serves no function, it is irrelevant to the Hillelites whether or not the farmer does so.[31]

5:8

A. Said R. Judah, "At first they would send [word] to the householders in the province [before Passover of the fourth and seventh years of the Sabbatical cycle saying,] 'Hurry to properly remove [agricultural gifts] from your produce (htqynw 't prwtykm) before (^{c}dšl') the time of removal arrives,'[32]

B. "until R. Aqiba came and taught that all produce which has not yet become liable to tithes (šl' b'w l^{c}wnt hm^{c}śrwt) is exempt from [the law of] removal."

M. 5:8

Judah's ruling assumes that we understand the following concerns. A crop is still growing in the field when the time for removal arrives. This crop contains in it produce which ultimately will be removed as heave-offering and tithes. At issue is whether or not we deem these agricultural gifts to be already in the crop. If we do, then they must be taken out before the removal. If these gifts remain in the crop, the whole field will have to be destroyed at the time of removal since it has in it produce which the farmer no longer may have in his possession. If, on the other hand, these gifts are not deemed to be in the crop, the law of removal does not apply to the produce at all. There is no need, in this case, for the farmer to separate agricultural gifts from the crop before the time of removal. With these facts in mind, we turn to the rules before us.

According to Judah, farmers were at one time required to separate from crops growing in the field at the time of removal all agricultural gifts which they owe from that crop. The supposition here is that these agricultural gifts to which the crop will become subject are already in the produce. If follows that if they are not taken out before the time of removal, the whole field will come under the law of removal.[33] Judah now informs us that Aqiba eventually modified this view (B). In Aqiba's view, agricultural gifts are deemed to be in only those plants which have matured. His reasoning is based on the fact that heave-offering and tithes are separated only from food which is ready to be eaten (M. Ma. 1:2-5). Plants still growing in the ground are not liable for the removal of these gifts and thus are not subject to the law of removal.[34]

5:9

A. One whose produce is unavailable to him (rḥwqym mmnw) [when the time for removal arrives]

B. must make an oral declaration [designating the required agricultural gifts and transferring them to their proper recipients].

C. M^{c}śh b: Rabban Gamaliel and the elders were traveling on a ship [when the time for removal occurred]. Said Rabban Gamaliel, "The tenth I intend to measure out [and designate as first tithe] is given to Joshua [who is a Levite], and the place [in which it is located] is rented to him. The other tenth which I intend to remove [and designate as poorman's tithe] is given to Aqiba ben Joseph, who will make it available to the poor (yzkh bw l^{c}nyym) and the place [in which it is located] is rented to him."

D. Said R. Joshua, "The tenth I intend to measure out [of the first tithe I will receive from Rabban Gamaliel and which I intend to designate as heave-offering of the tithe for the priest], and the place [in which it is located] is rented to him."

E. And they received rent-payment from one another.

M. 5:9
(C: b. Qid. 26a-b; b. B.M. 11a-b; y. Pe. 4:6)

The pericope consists of a rule, A-B, which is illustrated at C-E.[35] The time of removal now arrives before the farmer has taken heave-offering and tithes from his crop. Furthermore, he is unable physically to separate the required agricultural gifts from his produce, so that he is in danger of losing the entire crop. The point of A-B is that he may designate these gifts in his produce through an oral declaration (cf. M. Ter. 3:5) and may also transfer them in this way to the possession of their appropriate recipients, the priests, Levites and the poor. C-E supplies an example of how this might work. The owner designates the appropriate agricultural gifts in his produce through an oral declaration, and then transfers them to the appropriate recipients. In order to make the transfer valid, he then leases to the recipient the places in which the now-consecrated foods are stored, thereby giving the recipients legal access to the produce in question.

5:10-14

A. During the afternoon of the last festival-day [of Passover during the fourth and seventh years of the Sabbatical cycle, the farmers] would recite the confession [Deut. 26:13-15, stating that they have properly distributed or destroyed all consecrated produce from their domain, cf. M. 5:6].

B. What was the confession?

C. <u>I removed all holy [produce] from my house</u> [Deut. 26:13f] —

D. this [refers to] second tithe and [produce from] a planting's fourth year [of growth].

E. <u>I gave it to the Levite</u> —

F. this [refers to] the tithe for the Levites [i.e., first tithe].

G. <u>And I also gave it</u> —

H. this [refers to] heave-offering and heave-offering of the tithe.

I. <u>To the stranger, the orphan and the widow</u> —

J. this [refers to] poorman's tithe, gleanings, forgotten sheaves and what grows in the corner of the field [all of which are left for the poor];

K. even though [leaving] these is not a prerequisite for saying the confession (<u>š'nn m^{c}kbyn 't hwdwy</u>).

L. <u>From the house</u> —

M. this [refers to] dough-offering.

M. 5:10
(Sif. D. 303; H: y. Bik. 2:2)

N. <u>According to all the precepts you commanded me</u> —

O. lo, if he separated second tithe before [he separated] first [tithe, that is, out of order] he may not recite the confession.

P. <u>I did not transgress your precepts</u> —

Q. [this means] I did not separate [agricultural gifts] from one kind [of food] on behalf of another kind, and not from harvested produce on behalf of unharvested produce, and not from unharvested produce on behalf of harvested produce and not from new produce [i.e., produce harvested after the current <u>^{c}omer</u>] on behalf of old produce [i.e., from before the current <u>^{c}omer</u>] and not from old produce on behalf of new produce (= M. Ter. 1:5).

R. <u>And I did not forget anything</u> —

S. I did not forget to praise you and to mention your name [in connection with my crop].

M. 5:11
(Sif. D. 303; O: M. Dem. 1:4, Ter. 3:6, cf. T. Ter. 4:10, y. Dem. 7:6, Mekh. <u>mispha</u>ṭ 19; Q: M. Ter. 1:5; 2:4, b. Qid. 62a; Sif. Nu. 120; cf. T. Ter. 2:6-7, R.H. 1:9, b. R.H. 12b, b. Bek: 53b)

T. <u>I did not eat of it while in mourning</u> —

U. lo, if he ate [second tithe or produce of a planting's fourth year] while he was in a state of mourning before the burial (<u>b'nynh</u>), he may not recite the confession.

V. <u>Nor did I separate unclean produce from it</u> [as an agricultural gift] —

W. lo, if he separated it [the agricultural gift] when it was in a state of uncleanness, he may not recite the confession.

X. <u>And I did not give [any of its value] for the dead</u> —

Y. I did not [use its value] to buy a coffin and shrouds for the dead.

Z. <u>And I did not give it</u> —

AA. to other mourners [whose dead are unburied].

BB. <u>I obeyed the Lord my God</u> —

CC. I brought it to the chosen Sanctuary [i.e., Jerusalem] (<u>byt hbḥyrh</u>).

DD. <u>I did according to all you commanded me</u> —

EE. I was happy and made others happy [with the produce].

M. 5:12
(Sif. D. pis. 303; Y: y. M.S. 2:1; Shab. 9:4)

FF. <u>Look down from your holy dwelling-place in heaven</u> —

GG. We did what you required of us, now (<u>'p</u>) you do what you promised us.

HH. <u>Look down from your holy dwelling-place in heaven and bless your people Israel</u> —

II. with sons and daughters.

JJ. <u>And the earth which you gave us</u> —

KK. with dew and rain and with offspring of cattle.

LL. <u>As you vowed to our fathers [to give them] a land flowing</u>

with milk and honey —

MM. in order to give the fruit a [sweet] taste.

M. 5:13
(FF-HH: Sif. D. pis. 303)

NN. On the basis of this [verse] (LL) they said,

OO. "Israelites and mamzers [i.e., those of impaired lineage] may recite the confession,

PP. but not aliens or freed slaves, for these do not [hold] a portion of the land."

QQ. R. Meir says, "Also: priests and Levites do not [recite the confession] since they did not acquire a portion of the land."

RR. R. Yosé says, "[Priests and Levites do recite the confession for] they have the Levitical cities ([c]ry mgrš)."

M. 5:14
(NN-PP: Sif. D. pis. 301;
QQ: T. Bik. 1:2; RR: y.
Bik. 2:2)

Once the farmer has properly disposed of all consecrated produce in his possession (M. 5:6-9), he goes to Jerusalem and there recites the confession, averring that he has properly carried out this obligation. M. now supplies a phrase-by-phrase commentary on the text of the confession. This commentary reads into the text of the confession reference to the various agricultural gifts and how they are to be handled. NN-RR is an appended block of material which discusses who must say the confession, a point of law suggested by LL.

M.'s commentary to the confession makes three points, in logical order, as suggested by the Scriptural text itself. C-M, first, claim that the confession covers all agricultural gifts the farmer must separate. Thus each of the gifts known by M. is made to have an illusion in Scripture. N-EE discuss various laws as regards the proper separation and handling of consecrated produce. These laws, known elsewhere from M., are thus linked back to Scripture. FF-LL, finally, describe the blessings that result from proper removal of agricultural gifts, a fitting conclusion for this material. Only two points require special comment. The first is K. Its point is that gleanings, forgotten sheaves and what grows in the corner of a field are not deemed to be consecrated and thus are not liable to the law of removal. It follows that if the farmer has failed to leave these for the poor, he may still recite the confession. L is clearly out of phase since it repeats part of the verse already cited at C. It has been added here in order to complete C-J's list of consecrated gifts.

The interest of the final block of material, NN-RR, is in who is obligated to recite the confession. The basic rule is at LL, namely, that anyone who has a share in the Holy Land must recite the confession. Meir and Yosé (QQ-RR) disagree as to whether or not priests and Levites have a portion in the Land and are thus obligated for saying the confession. Meir claims that these have no allotment of land and thus are exempt from the obligation.[36] Yosé's argument is that the Levites do have possession of the Levitical cities (Nu. 35:2) and thus must say the confession.

A. R. Yosé says, "I removed all holy [produce] from my house —

B. this [refers to] dough-offering (M. 5:10L-M)."

T. 5:23 (p. 272, l. 65)

C. Another interpretation [is that] once you have removed it from the house you are not responsible for it [any longer].

D. The mouths of all those who fulfill the precepts are open in prayer to the omnipresent [and their prayers are answered] as it says [in Scripture], "[You shall pray to him and he will hear you, you shall pay your vows;] when you decide on a matter, it will be established for you . . . (Job 22:27f)";

E. and it says [in Scripture], "And Hezekiah turned his face toward the wall [and prayed to the Lord] saying, 'Please, God, remember [that I conducted myself before you in truth and with proper intention and did what you favor.' Then the word of the Lord came to Isaiah, 'Go say to Hezekiah, thus saith the Lord, the God of David your father: I have heard your prayers . . . '] (Is. 38:2-5)."

T. 5:24 (p. 273, ll. 66-68)

F. And it says [in Scripture], "Look down from your holy dwelling place . . . (Deut. 26:15)."

G. This is the place of looking down [referred to in the passage] "Truth will spring up from the ground and righteousness will look down from the heavens. [God will surely grant the good, our land will give her full yield] (Ps. 85:12)."

H. From the heavens — from the storehouse of good in the heavens, as it says [in Scripture], "May the Lord open for you his storehouse of good, the heavens, [to give the land rain in its season and to bless all your labors] (Deut. 28:12)."

T. 5:25 (p. 273, ll. 68-71)

I. And bless your people (Deut. 26:15) —

J. everything is included in this blessing, as it says [in Scripture], "Blessed are you in the city [and blessed are you in the field, blessed is the fruit of your womb, the fruit of your land, and the fruit of your cattle . . .] (Deut. 28:3)."

T. 5:26 (p. 273, ll. 71-72)

K. Israel —

L. all [these blessings come] through the merit of Israel, as it says [in Scripture], "Israel dwelt in safety, the fountain of Jacob alone, [in a land of grain and wine, even the heavens of which dripped with dew. Happy are you, Israel. Who can compare to you — a people saved by the Lord] (Deut. 33:28)?"

M. And the earth —

N. all [these blessings come] through the merit of the altar, as it says [in Scripture], "Make me an altar of earth [and offer upon it your burnt-offerings, your peace-offerings, your sheep and your cattle. For wherever I cause my name to be remembered, I will come to you and bless you] (Ex. 20:24)."

T. 5:27 (p. 273, ll. 72-75)

O. Which you gave us —

P. all [agricultural offerings] are [given] in response to [the number of your] gifts [to us], as it says [in Scripture], "Houses full of all good things [which you did not fill, dug out cisterns which you did not dig, vineyards and olive trees which you did not plant and you eat and are satisfied. Watch yourself lest you forget the Lord who brought out of Egypt . . .] (Deut. 6:11-12)."

T. 5:28 (p. 273, ll. 75-76)

Q. As you vowed —

R. this was the vow made to Abraham our father, as it says [in Scripture], "By myself I have sworn, says the Lord, because [of what you have done I will surely bless you and increase your progeny like the stars of the heaven] (Genesis 22:16)."

S. To our fathers —

T. all [these blessings come] through the merit of the tribes, as it says [in Scripture], "[Were you mad at the rivers, God, was your anger against the sea when you rode upon your horse, your chariot of victory? You stripped bare your bow from its sheath], the oaths of the tribes; say Selah (šb^{c}wt mṭwt 'mr slh), you gashed the earth with rivers (Hab. 3:9)."[37]

U. A land —

V. this is the land itself, as it says [in Scripture], "[These are the names of the tribes . . .] from the eastern side to the western side: Naphtali, one portion . . . (Ezek. 48:3)."

W. Flowing with milk and honey —

X. this teaches that the [giving of] tithes imparts flavor, aroma and plumpness (šmn) (add with E: and) abundance (dgn) to the crop.

T. 5:29 (p. 273, l. 76 —
p. 274, l. 80)

Y. R. Simeon b. Eleazar says, "[Failure to keep the produce in a state of] cleanness removed flavor and aroma [from the produce] and [failure to separate] tithes removed the plumpness and the abundance."

T. 5:30 (p. 274, ll. 80-81)
(M. Sot. 9:13; T. Sot. 15:2;
y. Sot. 9:15)

T. concludes with a commentary on the confession (Deut. 26:12-15), similar to that of M.'s. A-C refer to the opening phrase of the confession, assigning to Yosé the view given anonymously in M. 5:10L-M. T.'s own commentary, which begins at D, refers only to the final verse of the confession and deals with the blessing which will come upon Israel as a result of obedience to the laws of tithing.

The commentary unfolds in three thematic units, suggested by references in the text of the confession itself. D-J, first, declare that God in the heavens responds to the prayers of the faithful on earth who do his will. K-T then have the confession state that Israel also receives blessings through the accrued merit of its past. This merit accrues to Israel specifically on account of the people (K-L), the Temple (M-P) and the faithfulness of the ancestors (Q-T). Israel is thus assured of continued divine care despite the destruction of its cult. The third unit, U-Y, deals with the content of the blessing, as suggested by the reference to land in the confession. Israel will be blessed with health crops and abundanț harvests.

5:15

A. Yoḥanan the High priest [i.e., John Hyrcanus] did away with (1) [the recitation of] the confession concerning [the removal of] tithes (hwdyt hm^{c}śr).

B. Also: he dismissed (bṭl) (2) those who sing in the Temple the psalm of awakening (hm^{c}wrrym) [i.e., Psalm 94:24, "Awake, why do you sleep, O Lord . . ."], and (3) those who stun (hnwqpym) [the sacrificial animals before they are slaughtered].

C. (4) And until his time, hammers would pound [i.e., work was done] in Jerusalem [during the intermediate days of Passover and Sukkot].

D. (5) And in his time, no one had to ask [which agricultural gifts had been separated] from produce purchased from an ^{c}am ha'areṣ (^{c}l hdm'y).[38]

M. 5:15

(M. Sot. 9:10; A: y. Dem. 1:2; C: b. M.Q. 11a; cf. T. Sot. 13:9-10)

This is a list of five legal actions which are dated to the time of Yohanan.[39] The list, in fact, is a composite. B has been appended to A through the use of 'p. C-D shift the subject and themselves form a subunit balanced at the introductory formulas, <u>wᶜd ymyw/wbymyw</u>. The collection is located in our tractate because of A, which deals with the confession concerning the proper disposition of heave-offering and tithes. M. offers no clues, however, as to why these reforms were enacted. It is thus impossible for us to determine for certain the reasons for these reforms.

One explanation for these laws is provided by T. Sotah 13:9-10. T. takes as the result of a single reform both the abolition of the confession (A) and the fact that no one needed to ask about the status of food purchased from an <u>ᶜam ha'areṣ</u> (D). This reform required the farmers to separate from their produce only heave-offering, heave-offering of the tithe and second tithe. As a result of this action, the entire population came to adhere to a common set of tithing laws. Consequently, there was no need to inquire about the status of food purchased from an <u>ᶜam ha'areṣ</u>. On the other hand, the reform makes it impossible to recite the confession, since not all gifts are separated from the produce and distributed.

The singing of Psalm 94:24 was abolished, according to T., because of its suggestion that God sleeps, and therefore requires wakening. Stunning animals before slaughtering them on the altar (B) was halted because of the fear that the blow would blemish the sacrificial animal, rendering it unfit for the altar. The reference in C refers to labor being performed on the intermediate festival days of Passover and Sukkot. Yoḥanan allegedly ruled that work was not to be done in Jerusalem on these days, thus making the entire holiday period into a time of rest.[40]

NOTES

PREFACE

[1]My characterization of Tosefta as a supplement to Mishnah is based on my own work with these two tractates. The same conclusion has emerged from J. Neusner's literary analysis of the relationship between M. and T. in Purities XXI. His results are summarized in "Redaction, Formulation and Form: The Case of Mishnah," in JQR 70:3 (1980), pp. 14-16.

INTRODUCTION

[2]Besides second tithe, Israelite farmers are to separate two other offerings from the completed harvest. The first of these, heave-offering, belongs to the priests. It amounts to approximately two percent of the crop's total yield. A full ten percent of what remains, the so-called first tithe, is given to the Levites. From this produce the Levites themselves owe a portion to the priests. This is termed heave-offering of the tithe. The farmer owes in addition a second levy of ten percent. This is separated from the produce that remains after heave-offering and first tithe have been taken out. It is this levy which Mishnah calls second tithe. M.'s system of agricultural gifts is explained in detail in R. S. Sarason, A History of the Mishnaic Laws of Agriculture: A Study of Tractate Demai (Leiden, 1979), pp. 3-8.

[3]The notion that a set proportion of the farmer's crop is to be set aside for the support of the cult or its personnel is a common one in the ancient Near East. Israelite society in Biblical times naturally adopted this same system for the financial upkeep of its own cult. The best general introduction to Israel's system of Temple taxes is O. Eissfeldt, Erstlinge und Zehten im Alten Testament, a precis of which appears in RGG[3]. See also M. Weinfeld, "Tithes," EJ 15:1156-1162.

[4]Although Scripture's tithe is separated annually, it is eaten in Jerusalem by its owner only two out of every three years. Each third year, according to Deut. 14:27-29, the tithe is brought to the nearest city and there distributed to the unemployed Levites, the widows and the orphans as charity. Mishnah adopts the same calendar for its own tithe. Second tithe is separated during the first, second, fourth and fifth years of the Sabbatical cycle. The tithe set aside during the third and sixth for the poor is termed poorman's tithe. No tithes were separated during the seventh year since all agricultural labor was prohibited during that year.

[5]Modern exegetes have interpreted this passage to mean that the food must be eaten in some sort of sacred meal. See, for example, S. R. Driver, "Deuteronomy," in the ICC (New York, 1895); and H. H. Guthrie, "Tithe," in IDB, p. 634 (Nashville, 1962). M. differs in that it allows the farmer to eat second

tithe anywhere in Jerusalem and not only in the Temple compound. Nonetheless, M. holds that once the Temple was destroyed, second tithe no longer could be eaten. See, for example, M. 1:6-7, T. 3:13.

[6]Mishnah's authorities are interested in agricultural gifts in the first place because these are deemed to be holy. This is shown by the fact that the Order of Agriculture has tractates on all those gifts which are assumed to be in a consecrated status. Thus we have tractates on tithing in general (*maaserot*), heave-offering (*terumot*), second tithe, dough-offering (*ḥallah*), and firstfruits (*bikkurim*). There is no tractate dealing with first tithe, however. This is so because first tithe is not a consecrated offering, but simply a secular gift owed to the Levites. Mishnah's concept of holiness as it applies to agricultural gifts is the subject of a study by R. S. Sarason, "Mishnah and Scripture: Preliminary Observations on the Law of Tithing in *Seder Zera'im*," in W. S. Green, ed., *Approaches to Ancient Judaism II* (Missoula, 1980).

[7]*The Six Orders of Mishnah. I. The Order of Zera^{c}im* [Heb.] (Jerusalem and Tel Aviv: Bialik Institute and Devir, 1957), pp. 241-268.

[8]*The Mishnah with Variant Readings, Order Zera^{c}im* (Jerusalem: Makhon HaTalmud Ha Yisraeli HaShalem, 1972-75), II., pp. 247-313.

[9]*The Mishnah* (London: Oxford University Press, 1933), pp. 73-82.

[10]*The Babylonian Talmud: Seder Zera^{c}im, II. Maaser Sheni* (London: Soncino Press, 1948), pp. 279-311.

[11]*The Tosefta. I. Order Zera^{c}im* (New York: The Jewish Theological Seminary of America, 1955), pp. 243-274.

[12]The commentaries of both Sens and GRA can be found in the Romm edition of the Babylonian Talmud.

CHAPTER ONE

[1]M. does allow the farmer to sell such produce so as to transfer its status to coins. These coins are then taken to Jerusalem in place of the original and are used there to purchase other produce. This transaction is spelled out already in Deut. 16f. The action prohibited here, apparently, is the sale of such produce to another with the understanding that the purchaser acquires the food in its consecrated status and must eat it in Jerusalem. What is ruled out is for the farmer to sell to another his obligation to eat second tithe in Jerusalem. See Albeck and MB.

[2]The expressions *lhtn'wt* and *lht^{c}ṭr* appear in both MS traditions. Both terms are associated with money lending also in T. B.B. 4:2, where a third term, *lhtlmd* (= to practice) is unclear. Later exegetes offered several alternate readings: *lhtr'wt* for *lhtn'wt* and *lht^{c}śr* for *lht^{c}ṭr* (GRA). Maimonides

supplies lhtgdl for lht[c]ṭr in Yad 3:20. See Lieberman's discussion, p. 713.

[3]Ḥlwdh is traditionally translated as "rust." Technically, "corrosion" would probably be more accurate, as none of the commonly used metals for minting (copper, gold, silver, orichalcum) was particularly susceptible to rust. I have chosen to retain the more familiar translation.

[4]MB explains the rationale behind O-P as follows: When the first person says he is giving his wine away because he has no oil, he is simply stating a matter of fact, to explain why he is not giving his friend oil as well as wine. Thus, his friend is in no way obligated to give him oil in return, and the transaction is technically a gift.

[5]Ṣynyt is translated as pwdgrh in y. Shab. 8b. The Babylonian Talmud describes pwdgrh as a painful disease of the feet (Sotah 10a and Sanh. 48b). The first century B.C.E. physician, Celsus, describes a disease of the joints, called podagra, at some length in his De Medicina, Vol. I, pp. 455-461 and 463. Podagra is also discussed in Preuss, Biblisch-talmudische Medizin, p. 191. See also Lieberman's discussion, p. 715.

[6]Lichen is "an inflammatory skin disease with wide, flat papules, occurring in circumscribed patches and often very persistent" (Dorland's Illustrated Medical Dictionary). The first-century physician, Celsus, describes several lichen-like diseases in De Medicina, Vol. II, pp. 169-173. See also the discussion in Preuss, p. 398.

[7]It is unclear from the body of the pericope whether second tithe produce or second tithe coins are under discussion. Both interpretations are possible. Maimonides in Yad, M.S. 3:16, assumes the subject to be oil (as in T. Terumot 9:12, M. Shebi[c]it 6:4 and M. Shabbat 12:12. HD understands the prohibition to refer to coins as is the case in M. Shabbat 6:6.

[8]The question arises in M. 2:1-4 as to whether or not anointing with oil is deemed to be a form of consumption and thus permissible in Jerusalem. T. here seems to claim that anointing with oil in the status of second tithe is not permissible. It is possible, of course, that T. here is concerned with the use of other kinds of consecrated produce to make medicines or balms.

[9]The word, 'symwn, denotes a coin that has no image, or is so poorly minted that its value is in question. Exegetes derive the requirement to use only properly minted coins for the redemption of tithe from Deut. 14:25 — wṣrt hksp — that is, the money must have an image — ṣwrh. See Lieberman's discussion on p. 715.

[10]As we have stated, certain parts of an unblemished firstling, must be offered on the altar before the meat can be eaten. It seems odd, then, that M. would allow the priest to sell an unblemished firstling to a non-priest, since this person will use the animal for secular purposes. By selling the beast to a non-priest for consumption, then, the priest is depriving the altar of its due. In order to account for M.'s law, Maimonides declares that while a non-priest may purchase the animal, he may

not slaughter it while it is still in its unblemished state. Rather, he must guard the animal until a blemish naturally appears in it. When this occurs, the animal is in all events unfit for the altar and available to non-priests, as E(2) makes clear. Now the purchaser may freely use the meat of the firstling without fear that he has improperly used what belongs to the altar.

[11]Exegetes offer several opinions as to the function of these coins. MB suggests that they aided the bathhouse attendants in keeping track of the number of patrons. They may also have served as tokens signifying prepayment. See Lieberman's discussion, p. 715.

[12]HD, basing his interpretation on the y. M.S. 1:2, interprets this passage to ban any coins from a revolt like Bar Kokhba's. Since Jerusalem will only be legitimately conquered at the return of King David, coins of any other regime are not valid.

[13]According to Bunte, p. 151, "money of Jerusalem" refers to the coins minted under the leaders of the rebellion during the reign of Nero (!). His point is that the two types of coins mentioned in T. are equivalent in that both were minted by revolutionaries. On the other hand, Lieberman says that at issue are the coins of ancient Jerusalem, whose value was determined by weight and not by their minted design. This exegesis attempts to harmonize T.'s "Jerusalem coins" with the 'symwn of M. It is conceivable that "money of Jerusalem" may be another name for the coins minted by Bar Kokhba.

[14]The Soncino translation to B.Q. 98a (Nezikin I., p. 569) takes qṣṭr' (and several variant spellings) to be a Hebrew transliteration for the Latin word denoting military camp and thus not a reference to one specific place. The Encyclopedia of Palestine (IV: 831, 833) identifies several sites whose names are formed from the word "castra." I have chosen to translate the phrase using the indeterminate "a Roman camp" rather than a specific place name as better serving the sense of T. All Roman camps would be equally off-limits to Jews after the Bar Kokhba revolt.

[15]The place name, hr hmlk, is a matter of some disagreement. Soncino translates the phrase as "King's Mountain" — an unidentified site in Palestine. Bunte, p. 152, cites our T. (as "Bet-ha-melek") and says that the name refers to a place far from Jerusalem. Press, The Encyclopedia of Palestine (II: 208), however, identifies hr hmlk with the mountains of Judah in the vicinity of Jerusalem. This is also the opinion of M. Avi-Yonah in his article under the heading "Har Ha-Melek" in the Encyclopedia Judaica (VII: 1337). I have adopted this later opinion for my translation. Subsequent to the revolt of Bar Kokhba, the Roman authorities turned Jerusalem into a Roman polis and banned all Jews from entering. Thus it was appropriate to use the Jerusalem area as an example of a place that was inaccessible to Jews. The original location of the site was apparently forgotten by Amoraic times (see b. Gittin 57a, for example).

[16]GRA, following: b. B.Q. 115b emends to: hry pyrwt šyš ly btwk byty mḥwllyn ^{c}l m^{c}wt hllw. Lieberman suggests the following reading: mḥwllyn ^{c}l[yhn] m^{c}wt šyš ly btwk byty.

[17]The term bśr t'wh, literally "meat of desire" is derived from Deut. 12:20: "You may eat as much flesh as you desire (t'wh npšk l'kl bśr)." In M. the phrase is used to designate meat bought for private consumption, as opposed to meat purchased for the purpose of an offering.

[18]Most MSS indicate a passive mood, which is clearly required by the text.

[19]This type of cheap wine was widely known in the Roman Empire. It was made by steeping stalks, husks and skins of grapes in water until the mixture began to ferment. Pliny discusses the manufacture of this wine, termed "worker's wine," in NH XIV: 10. See also Dalman, AS. VI, p. 371.

[20]Y. M.S. 1:3 claims that originally domesticated animals could be bought with consecrated coins for use as second tithe. Sages later prohibited this practice. The reason, y. claims, is that so many domesticated animals were purchased for private use that there were none left for peace-offering. In order to insure a steady supply of peace-offerings for the altar, they prohibited domesticated animals from being purchased for other purposes. I do not know where y. gets its information, and so I cannot evaluate its claim.

[21]The spelling of this place-name varies considerably. D gives 'klym, A has 'yblyys, T. Niddah 9:18 offers kfr 'blyn. Press, The Encyclopedia of Palestine I, p. 4, identifies the village as "Avelin" located in northern Israel. It was the site of the Crusader fortress of Abelin.

[22]Neusner, The Tosefta, IV, p. 233-234. I have changed Neusner's "meat of desire" to "ordinary meat" to maintain consistency with my translation.

[23]M. Parah 11:5 takes for granted that one requiring immersion by the rabbis is prohibited from eating tithes. All exegetes offer the eating of unclean food as an example of how such uncleanness is acquired. The tithe prohibited here is generally understood to be second tithe. See MS to this mishnah. Also TYT and TYY to M. Parah 11:4, which parallels 11:5's discussion.

[24]Mwrys is a dish made of chopped fish in salt water or wine.

[25]D: rytk'wt; A: dwrk'wt; Lieberman (p. 721) suggests reading rwkb'wt or rykb'wt, which he translates as the fan-shaped or broom-like branches of the date-palm. Jastrow, p. 290, defines dwrk'wt as pomace.

[26]The stalks, husks, etc. of grapes or olives would be pressed in order to extract their juice. The resulting mash was termed gpt. Loew discusses the wine made from these juices, Flora I, 96, and the use of the mash as fuel, II, 290. Instead of being pressed, the refuse may be soaked in water to produce the temed-wine mentioned in M. 1:3H.

[27]Salt and water are explicitly excluded from the types of edibles which can be used to make an erub (M. Er. 3:1). The point is that for these purposes, the householder may use only proper foods.

[28]Maimonides wants to relate the rules of these pericopae to Scripture. He cites Deut. 14:26, which states, "and spend the money [consecrated as second tithe] on anything you desire [such as] cattle, sheep, wine or intoxicating beverages . . . ." From Scripture's data Maimonides concludes that items which the farmer may purchase as second tithe must be agricultural produce and must not be attached to the ground. On the basis of these criteria, he explains M.'s exclusion of water and salt, which are not agricultural produce, and of plants still growing in the ground, since they are not separated from the earth. Although Maimonides does account for the data, his explanation is over-complicated. Mishnah is concerned simply that the farmer purchase as second tithe items which are in the category of food. Salt and water are excluded because they provide no nourishment, and plants still in the earth are not ready for use as food.

[29]The outcome, according to the Shammaite opinion, is parallel to that of a case in which a farmer dedicates an animal to the Temple and then attempts to name a substitute for it. The original animal is not deconsecrated, but the substitute nonetheless enters the status of dedication. The basis of this law is Lev. 27:10. Cf. Neusner, Holy Things, IV, pp. 87-88.

[30]Mespilus Monogyna. A common Palestinian bush or tree. The sprouts here probably refer to the young, prickly branches (Loew, Flora, III, p. 255).

[31]Arum Colocasia. Loew, Flora, I, pp. 214-226, identifies Mishnaic luf with A. Hygrophilium and A. Palaestinium.

[32]Carthamus Tinctorius. (Loew, Flora, I, p. 394ff). The seeds of this plant are edible and thus may be acquired with second tithe coins. Since safflower is not planted for its seeds, however, they do not become unclean as food. So Lieberman, p. 725.

[33]So Loew, Flora, I, p. 396. HD identifies hlwt hry[c] with kwrkwm (saffron) mentioned in T. 1:14. The phrase is translated as "saffron-seed cakes" in JE, III, p. 334. Lieberman (p. 725) does not identify the item beyond saying that it was named after its distinctive shape.

[34]"The white heart or terminal bud of a palm (cabbage tree) used as food," (Jastrow, p. 1340). MB explains that when the qwr is soft, it may be eaten when cooked. It does not become susceptible to uncleanness as food until it has been properly prepared. Loew wants to identify qwr as a kind of gourd (Flora, I, p. 545).

[35]MB explains that these unripe berries are similar to palm tree pith in that they need to be sweetened before being eaten. Unlike qwr, however, they are easily prepared by being placed in a warm oven (Loew, Flora, II, p. 342).

[36]That is, one says the blessing said over fruit when one eats the qwr. They are like trees in that they do not contract uncleanness as food (so Albeck to M. Uq. 7:3).

[37]They are exempt from tithes because they are not fully ripe (Albeck, to M. Uq. 7:3).

[38]Crocus satinus. (Loew, Flora, II, p. 7f).

[39]According to Lev. 15:14 and 29, men or women who suffered a flux remain unclean for seven days. On the eighth day, they bring two turtledoves or two pigeons to the altar to complete their purification process. A similar rule holds for a woman who has just given birth. When her period of uncleanness is over, she brings a lamb and a turtledove or pigeon to the Temple. If she is poor, she may bring a second turtledove or pigeon in place of the lamb (Lev. 12:8). One of the birds is designated as a sin-offering and the other becomes a burnt-offering. In either case, the animal is slaughtered, its blood is sprinkled on the altar and its meat is consumed by the priest.

CHAPTER TWO

[1]Allium Porrum. Also referred to as krty. See Loew, AP, pp. 226-228.

[2]I owe this interpretation of the pericope to Martin Jaffee. Classical exegetes have read this pericope atomistically and thus have failed to note that A-C and D-H make the same point. They have thus interpreted these two blocks of material independently of each other. As to A-C, Maimonides, Bert., TYY and Albeck take the law to be that we do not force a farmer to eat consecrated food which has spoiled. The point, according to these commentators, is that since this food is no longer deemed a proper food, the farmer need not treat it as such and so may discard it as he would unconsecrated garbage. TYY adds that he also is not required to eat raw food that is normally cooked. We must reject this interpretation, however, since it has no basis in M.'s language and is in fact contradicted at B-C.

The concern of D-H, according to Bert, TYY and Albeck, is that consecrated produce may not be wasted. M. thus prohibits adding spices to oil because some of the oil will be absorbed by the herbs and will thus be lost. Although this rationale is reasonable, it should apply as well to wine, which also is absorbed by added herbs. Since this is not a consideration in the case of wine, I do not see why it should be in the case of oil. On this basis, I have rejected their explanation. Further, by claiming that the point is that the added spices change the use of the oil, but not of the wine, I keep my explanation in line with the theme set forth in A-C.

[3]There is some disagreement as to the meaning of the term "perceptible" in O-P. Maimonides, followed by Sens, Bert., TYY and TYT, interprets the rule to mean that the added ingredients must be physically present in the increased bulk of the mixture. Using the firewood used to heat the oven as an example, Maimonides explains that the wood is not allotted any value because it is not physically present in the finished loaves. MR rejects this interpretation, reading O-P to require simply an increase in bulk (or flavor) in order for the increased value to be divided proportionately. He thus must explain the ruling in M-N on other grounds. He prefers to make a distinction between an agent (gwrm) and a flavoring. Only a flavoring is assigned a portion of the increased value in the case of increased size. Since Maimonides' firewood is not a flavoring,

it is not given any value even though the dough is certainly expanded in size during baking.

4"A drink or brew made of pounded groats and spices, a spiced drink" (Jastrow, p. 553). B. Ber. 38a indicates that ṭrym' may be made of sesame, safflower or fruit kernels.

5It is possible to interpret C to mean that consecrated food may be processed even in unusual ways as long as none of it is wasted. The fruit in A-B will be discarded as soon as it has sufficiently flavored the water. The entire bulk of the puree will be eaten, however. It is thus a licit use of second tithe, even if what is normally eaten is now made into a drink.

6Also 'lwnṭyt (Jastrow, p. 68) and 'lwntyt (Aruch, I, p. 100). The word is a corruption of the Latin oenantius. Pliny describes the use of a lotion called "vinum oenanthius" in NH xii:132. The use of 'lntyt as a kind of curative balm appears in several rabbinic sources (T. Shab. 12:12, y. Shab. 14:3). According to b. Shab. 140a, this mixture is made up of old wine, pure water and balsam and was prepared for use in the bathhouse. The opinion of the Gemara that the mixture was drunk after emerging from the water is probably erroneous, since all other sources agree that it was an unguent.

7Viciae Ervilia. Loew, AP, P. 228.

8Trigonella foenum grecum. Loew, AP, pp. 316-317.

9Jastrow, p. 1287, defines ṣmḥwn as the soft seed taken from the pod. Maimonides, Albeck, Bert. and TYY interpret M. to allow fenugreek declared second tithe to be eaten only when the plant is first sprouting. This interpretation is inadequate. We know that fenugreek, as well as vetches, remain digestible even after the plant has matured and the seeds become hard. M. 2:3-4 explicitly state that once the seeds are hard, all the farmer needs to do is soak them in order to make them edible again. The fact that the seeds of the mature plant must be prepared before they can be eaten does not relieve the farmer of the obligation to eat such plants declared second tithe in Jerusalem. Rather, M. establishes the time at which fenugreek, or vetches, become edible and enter into the category of food subject to the laws of second tithe. The logic of the law here throws light on the interpretation of M. Ma. 1:3. That pericope is generally understood to mean that the plant becomes liable to tithes as soon as its seeds are mature enough to sprout on their own. Yalon reflects this interpretation in his vocalization of tṣmḥ in the pi'el. If we read the verb in the qal, "when the plant sprouts," the time of eligibility for tithes specified here will coincide with the time the plant becomes edible according to our pericope.

10My translation follows Jastrow, p. 1539. Krauss suggests "cut" ("schneiden"), II, p. 130. Danby, p. 75, and Soncino, p. 209, give "rub." The idea is that the vetch is worked in such a way that the shell around the seed is cracked, making the seed more digestible. "Crush" seems to convey this sense best.

11Maimonides offers another interpretation of Simeon's rule. He is concerned that the consecrated oil will rub off onto the hands of the masseur. In such a case, a person other than the owner will have derived benefit from the oil, in violation of the laws of second tithe. Albeck argues that the farmer might leave Jerusalem with unabsorbed oil still on his skin. Should the farmer incidentally brush up against someone, consecrated oil will have been consumed outside of Jerusalem, again in violation of the laws of second tithe. To avoid this possibility, Albeck has Simeon ban anointing with oil designated as second tithe altogether.

[12]The Houses' dispute here reflects a basic philosophical disagreement over the criteria by which one determines the primary status of an object which has several possible uses. The view adopted by the Shammaites is that we treat an item according to the use which involves the most restrictions. In this way, we are in all events sure not to transgress the law. Food is susceptible to uncleanness, for example, and therefore must be handled differently than are non-foods so that it does not become unclean. Since fenugreek may be used as a food, it follows that Shammaites treat this plant as if it were a food and so preserve it in a state of cleanness. The Hillelite position is that an item is categorized according to its most common use. Since fenugreek is most commonly used as a shampoo, the Hillelites will assign fenugreek the primary status of a non-food. This same issue appears in M. Toh. 8:6. See Neusner, Purities, XI, p. 193f.

[13]This rule is based on Scripture's statement that consecrated produce may be sold only if the farmer is unable to transport the produce itself to the city. It follows that once the farmer has brought the produce inside the city, he forfeits his option to sell it. This remains so even if the farmer subsequently takes the consecrated food out of the city again. See my discussion below in connection with M. 3:5-6.

[14]An egg's bulk is the standard measure used by M.'s formulators for the minimum volume of unclean food which can convey uncleanness to food with which it comes into contact. See, for example, M. Toh. 3:4.

[15]Cf. M. M.S. 3:10.

[16]The question whether or not a priest may feed his heave-offering to cattle is never explicitly discussed in M. Our pericope, as well as M. Ter. 11:9, does allow vetches declared heave-offering to be used as cattle fodder. The implications of this rule are read in two ways. Albeck, MS., Sens, TYT, and TYY claim that the law here makes a specific exception for vetches and that heave-offering may not normally be used to feed cattle. Maimonides takes the opposing view, that M. here specifically includes vetches within the general rule that heave-offering is available to the priest for use as fodder. The logic of the law supports Maimonides' position. Heave-offerings are given to the priest in order that he may support himself and his household. Since the priest's cattle are dependent upon him for their sustenance, it follows that the priest may feed them from his heave-offering income.

[17]The susceptibility of vetches to uncleanness as a food is debated by Judah and Simeon in T. Uq. 3:13. Judah there agrees with our Shammai that vetches are susceptible to unclean-

ness. Simeon holds the opinion ascribed to Aqiba in our pericope. See also T. Uq. 3:14.

[18]Pliny, NH, XXIV:187, describes the use of fenugreek as a shampoo: ". . . the meal with wine and soda quickly removes scurf and dandruff on the head."

[19]Lieberman, TK, p. 249.

[20]TYT attempts to reconcile the two opposing views found in M. In his view, the preferred method is that outlined in M. 2:5. The farmer picks coins from the mixture and designates them as second tithe in place of the lost money. In order to carry out the procedure properly, the farmer must declare that if the coins he selects are already consecrated, they are to remain so, and if they are unconsecrated, they are to take on the status of second tithe from the lost coins (cf. y. M.S. 2:6). It is likely, however, that the farmer will forget to make this declaration if he has to pick out only one coin. To make sure that the status is transferred properly if only two coins are involved, therefore, he must use the longer method described in M. 2:6.

[21]Holiness is conceived of in Mishnah as a kind of physical quality which is uniformly dispersed throughout a collectivity. It is concentrated and localized in a certain portion of the batch by the owner's oral declaration. This charged portion is then separated as the appropriate consecrated gift. I owe this description of the rabbis' conception of holiness to Richard Sarason. His inclusions are presented in detail in a paper entitled, "Mishnah and Scripture: Preliminary Observations on the Law of Tithing in Seder Zera^{c}im," delivered at the Fourth Max Richter Conversation in Providence, R.I., June 1978, to be published in W. S. Green, Approaches to Ancient Judaism II (Missoula, Scholars Press, 1980).

[22]The law here reflects the Greek conception that all items have two values, their intrinsic, material value and their market, or utility, value. This view derives from Plato and is adopted by Aristotle (cf. Politics, Bk. I, Chap. 9). This conception yields the result that a coin has two different values. On the one hand there is an intrinsic worth in its metal and, on the other hand, it has a particular face value. A dinar in silver coins and a gold dinar thus have equivalent face values but may have different value in terms of their material. The point of the law is that if the status of second tithe is being transferred from one set of coins to another, the receiving coins must have a higher intrinsic worth than the original coins, even though the face value of the two sets is identical.

[23]The law refers to items which are dedicated for the use of the Temple. These items are designated either as an offering or as a source of income for the maintenance and upkeep of Temple property. Since dedicated items belong to the Temple and are used solely for its benefit, they are deemed to be holy, and in this regard are analogous to produce consecrated as second tithe. Like second tithe, dedicated items can be deconsecrated by transferring this status to coins. As far as the law here is concerned, dedicated items are exactly parallel to items designated as second tithe.

[24]According to Lieberman, who reads this essay in light of b. B.M. 44b, the discussion revolves around the relationship of coins versus commodities (pry). That is, one metal may be deconsecrated by another only if the first metal is deemed a commodity in relation to the second. Thus, for example, gold may deconsecrate silver because the transaction is analogous to a coin, in this case gold, acquiring a commodity, silver. Silver, however, cannot deconsecrate gold because commodities do not purchase coins. If we apply this logic to the debate in our pericope, the dispute works out as follows: R. Eleazar considers copper to be a coin in relation to all other metals because copper coins are extremely common. Thus copper can deconsecrate silver and gold. Rabbi deems gold a coin in relation to copper and silver because of its greater desirability. Consequently, gold may not be deconsecrated by other metals, which are commodities in relation to it. Both gold and copper are coins in relation to silver, allowing silver to be deconsecrated by either metal.

[25]See also M. Dem. 1:2, Sarason, Demai.

[26]In Arak. 8:1 and Meilah 6:2, M. specifies that dedicated items are deconsecrated with money or that which is equivalent to money (šwh lksp). The status of dedication, then, can be passed to a wider variety of items than can the status of second tithe, which may be transferred only to money.

[27]M. does not spell out who these "disputants before the sages" are. Later tradition has given them an identity, however, in an effort to supply background for an understanding of how the Tannaim developed their law. Sanh. 17b claims that these disputants are Simeon b. Azzai, Simeon b. Zoma, Ḥanan the Egyptian and Ḥananian b. Hakinai. In the same place, Naḥman b. Isaac adds to the list the name of Simeon the Yemenite. Since we do not know the source of the Talmud's tradition, we have no way of evaluating its truth. Bacher, Die Agada der Tannaiten (Strassling, 1903), p. 406f, nonetheless, accepts the Talmud's account and goes on to claim that these were young scholars, not yet ordained, who were allowed to take part in some of the legal discussions of the academy. It is not clear to me how Bacher arrived at this conclusion, since the evidence he cites, even if it were reliable, does not support his claim.

[28]Alternatively, it is possible to read the Meir-sages dispute as a gloss to the Hillelite position. The Hillelites have told us that different items of the same genus, such as copper and silver coins, may be treated as an integrated unit. Now Meir may simply be arguing that this is not true for a collection of items of different genera: produce and coins. That is, such a collection, in his view, may not be treated as an integrated unit. Sages, by disagreeing, extend the Hillelite view by claiming that any items collected together for a single purpose may be deemed to form a single unit.

[29]Bert. interprets these two pericopae in light of his conception of the monetary situation which would be created in Jerusalem if the law were according to the Shammaites. In B, they ruled that the farmer may exchange only copper coins for silver. According to Bert., this would allow the farmer to exchange all his copper coins for silver. Pilgrims to Jeru-

salem, then, would bring with them only silver coins. This would create a glut of silver and a corresponding shortage of copper in Jerusalem, thereby reducing the value of the pilgrims' consecrated money. The Hillelite view, he holds, is tailored to avoid such a situation by making it difficult for the farmer to convert all his copper to silver, since, according to Bert., he can exchange only as much copper as he has small silver coins. Consequently, he will bring copper coins as well as silver coins with him to Jerusalem. Once in Jerusalem, the farmer exchanges the silver for smaller coins. If he receives only copper coins in return, as the Shammaites require, there is a danger, claims Bert., that the coins will rust before the farmer has a chance to spend them. Should the farmer choose to exchange the left-over copper coins for silver, he would then have to pay a fee to the money-changer, an illegal use of second tithe. Thus he receives both silver and copper to begin with when he changes consecrated silver selaᶜs. RABAD to M. Ed. 1:9-10 approaches these pericopae as I have done by considering the relative intrinsic values of copper and silver.

[30]The idea that produce may be sanctified on the basis of a future event also appears in M. Ṭebul Yom 4:4. In this pericope, the farmer has tithes from which heave-offering has not been removed. He now removes a portion of the produce and places it in a jug that is unclean in the status of ṭebul yom, which will become clean at sundown. The farmer stipulates that the produce in the unclean jar is to be considered consecrated as heave-offering as soon as the sun sets and the jug is clean. When the sun sets, the designated produce enters into the status of heave-offering in accordance with the farmer's declaration. See Neusner, Purities, Part XIX, pp. 55-58.

CHAPTER THREE

[1]In its present form, the pericope seems to be talking about produce in general. As such, it prohibits the farmer from offering as payment for work any produce he wants taken to Jerusalem. Nothing in M. prepares us for such a prohibition on the use of ordinary produce. If, however, we assume that our pericope has in mind produce in the status of second tithe, its prohibition becomes understandable. Now the force of the pericope is to limit the use the farmer makes of consecrated produce which he must consume in Jerusalem. The law, then, is in agreement with the laws of M. 1:1, which prohibit the farmer from deriving any benefit from produce in the status of second tithe.

[2]MR understands this pericope to refer to a farmer who is of priestly status. While he himself can use heave-offering as second tithe, he is not able to give his second tithe away to commoners. There is thus a restriction on the way he may use second tithe. For this reason, MR claims M. prohibits him from purchasing heave-offering for use as second tithe. MR's interpretation clearly is influenced by M. 3:1, which declares that second tithe may be given away. MR wants to carry this notion forward, declaring that the farmer in fact must be able to give his second tithe away. I see no reason for making this claim. In addition, MR supposes that the text has in mind a farmer who

is also a priest. At no point, of which I am aware, however, does M. assume that the householders it legislates for are priests.

[3]The laws of refuse (piggul) and remnant (notar) apply to the flesh of animals which are sacrificed on the altar. The flesh is deemed to be refuse if, while the priest was sacrificing it, he formed the notion of eating it at an improper time. The flesh is considered a remnant if it is not consumed within two days after the sacrifice is carried out. The laws of uncleanness, of course, apply to heave-offering as well as to peace-offerings. Heave-offering, however, only becomes unfit if it comes into contact with food unclean in the first, second or third remove. Peace-offerings are rendered unfit even if they come in contact with food unclean in the fourth remove. Peace-offerings are therefore more susceptible to uncleanness than are heave-offerings, which is the point of the law in our pericope. References to these three modes of rendering consecrated items unfit as food often appear together in M. See, for example, M. Toh. 3:4, M. Ḥul. 8:6, M. Tem. 7:1, M. Me. 1:2-3, 4; 2:1-9; 4:1, M. Men. 11:8 and M. Makkot 3:2.

[4]Generally dema'i refers to produce which might not have had tithes removed. One who is meticulous about the tithing laws must remove first and second tithe before eating the produce in order to insure that these offerings have been separated at least once. If tithes had already been separated, however, what he now takes out is not a tithe. Whatever is separated, then, can be only of doubtful status. If second tithe removed in this way is sold, the coins received in exchange are also of doubtful status. It is to such coins that the law of the pericope refers.

[5]Sens, Bert., MR and TYY want to harmonize Simeon in C with the ruling of B. They claim that everyone agrees that once a particular batch of produce has been brought into Jerusalem, that particular produce must be consumed there. Simeon b. Gamaliel's position is taken to mean only that the produce may be taken out of the city if the farmer intends to process it by grinding or baking it for example. Once the processing has been accomplished, Simeon would agree that the produce must be returned to the city for consumption. This interpretation is possible only if we read into C matters which have no basis in the text.

[6]The notion that produce is liable to the removal of heave-offering and tithes as soon as it is ready for consumption is found in M. Ma. Chapter One. The point is that tithes are due only from produce which can be used as human food. If produce is edible in its raw or natural state, it is subject to the removal of tithes and heave-offering as soon as it is harvested. If, however, the produce must be processed in some way before it is edible, it does not become liable to the separation of these gifts until such processing is completed. Once liable, the produce may not be eaten by common farmers until all agricultural gifts are removed. See also the introduction to Jaffee, Maaserot.

[7]It seems odd that a question discussed by the Houses should still be a subject of intense controversy in Usha, as

M.3:5 (Simeon b. Gamaliel) claims. This, added to the assertion that the Houses' opinions themselves were a matter of disagreement in the time of Rabbi, suggests that the debate is in fact of Ushan vintage, pseudepigraphically attributed to the ancient Houses.

[8]M. Ma. 1:1 rules that produce is liable to the removal of heave-offering and tithes only if it is edible, grows from the soil and is owned by the farmer. Any plant that does not fulfill these three criteria is not liable to the removal of tithes. It follows that if the farmer declares part of his crop to be ownerless, that portion will not become liable to the removal of tithes. See Jaffee, Maaserot.

[9]See, for example, M. Ter. 1:9, 2:1.

[10]It is not clear from M.'s language what the position of the consecrated fruit is to the bough. I have assumed that the law refers to fruit growing on the bough since this seems to be the most natural explanation. Medieval commentators, however, have explored other possibilities. Maimonides holds that the law refers to the status of the ground overshadowed by the limb. That is, a farmer who has consecrated produce in his hand and who walks under the bough is deemed to have entered the territory of Jerusalem. Accordingly, the fruit may no longer be sold. Read in this way, our pericope is parallel to M. Neg. 13:7 which declares that a clean person walking under a tree overshadowing an unclean person is deemed to enter the unclean person's domain and is himself rendered unclean. Another possibility is explored by TYY. He suggests that the law refers to the status of fruit held by a farmer sitting on the limb. His point is that since the tree is growing in the city, all of its branches are considered to be extensions of the city.

[11]M. Ma. 3:10 discusses a similar case. There we have a tree growing in the Land of Israel but with a bough extending over the border. The question is whether fruit growing on this limb is deemed to be within the Land, and thus subject to the removal of tithes, or whether it is outside the Land. According to M. Ma. 3:10, all of the fruit growing on the tree is judged to be growing in the territory in which the tree's roots are located. The actual location of the fruit itself, inside or outside the border, is not deemed to be important. I do not consider this law to be contradictory to ours. M. Ma. is concerned with the source of the produce's nourishment since only produce growing in the Land of Israel is liable to the removal of tithes. Thus the location of the roots is probative. Our pericope, to the contrary, is concerned with the actual physical location of the produce. The location of other parts of the tree is immaterial.

[12]I follow Lieberman's emendations of GG-JJ. See his comments, TK, p. 743.

[13]For the laws governing the location in which Most Holy Things and Lesser Holy Things may be slaughtered and eaten, see M. Zeb. Chapter 5.

[14]See M. Arak. 9:5; T. Arak. 5:14.

[15]This pericope has puzzled classical commentators, all of whom explain it on the basis of notions not supplied by M. Sens Bert., MR and TYY attempt to explain the pericope on the basis of y. M.S. 3:8, which assumes that the rules for uncleanness derived from a Father of uncleanness are based on Scripture, while the rules governing uncleanness derived from an Offspring of uncleanness are rabbinic. Produce in the status of second tithe which is rendered unclean by a Father, therefore, is deconsecrated on Scriptural authority and so may be removed from the city. If the uncleanness is derived from an Offspring, however, it is deconsecrated only on authority of sages and must therefore remain in the city. Since M. does not know the distinction y. makes between the authority of Scripture and that of the sages, it is clear that this explanation is unacceptable. Maimonides offers an alternative view, arguing that the status of second tithe inhering in produce is diminished in proportion to the level of uncleanness the produce contacts. Produce rendered unclean by a Father loses all of its sanctity and so may be taken out of the city, while produce rendered unclean by an Offspring loses only part of its status as second tithe and so must remain in Jerusalem. Since M. at no time claims that the level of sanctity in an item is in inverse proportion to its state of uncleanness, we must reject Maimonides' view as well.

[16]It is clear that B-C are later glosses and not part of the original protasis. They lead us to believe that the level of uncleanness and the location of the produce when it is rendered unclean should not figure in the Houses' dispute which follows. The Houses, however, make distinctions based precisely on these considerations. C, which declares that the uncleanness is contacted either inside or outside Jerusalem also stands in contradiction to A, which states that the produce is rendered unclean only after it enters Jerusalem. The protasis becomes more intelligible when we read B-C as later glosses, drawn into our pericope on account of the distinctions made in E and G. Although B-C intend to clarify the superscription by giving all four possibilities referred to in the Houses' dispute, they in fact only confuse matters. For a further discussion of the substantive problems posed by this pericope and a more detailed attempt to reconstruct its tradental history, see Neusner, Pharisees, II, pp. 103-104.

[17]A Father is a primary source of uncleanness, such as an insect or a corpse. A susceptible item which comes into contact with a Father of uncleanness itself becomes unclean and is termed an Offspring. It is capable in turn of contaminating foods and liquids with which it comes into contact. These foods are deemed to be in the second remove of uncleanness. They no longer have the power to contaminate unconsecrated foods. Heave-offering and Holy Things which come into contact with an item in the second remove, however, do become unfit for eating. See M. Toh. 2:3-7. For a fuller discussion of the implications of various levels of uncleanness, see also Neusner, Purities, XI, pp. 55ff.

[18]A similar concern with taking out of a sanctified area items rendered unclean by a major source of uncleanness is seen in M. Sheq. 8:4. Here the Temple veil has become unclean. If the veil became unclean by contact with an Offspring of unclean-

ness, it is immersed inside the courtyard. If it was rendered unclean by contact with a Father of uncleanness, however, it must be taken out of the courtyard to be immersed and dried.

[19]The words "that which was rendered . . ." in line E through "except for" in G have been added by Lieberman on the basis of A.

[20]The words "Let it be redeemed" in J through "Offspring of uncleanness" at K have been added by Lieberman on the basis of A.

[21]Eleazar's ruling should be compared to the similar position articulated in M. Sheq. 8:4. Cf. note 18 to M. 3:9 above, p. 207.

[22]Both authorities in I-L are interested in establishing criteria according to which unclean produce may be removed from the holy city. It is possible that the Houses' opinions as we know them from M. are in fact a conflation of the views of these two authorities. Eleazar allows the farmer to remove all virulent unclean produce while Aqiba allows him to remove all produce rendered unclean outside the city. They agree that produce rendered mildly unclean in the city must remain. This is precisely the view ascribed to the Hillelites in M. 3:9F-G. Conversely, Eleazar requires mildly unclean produce to remain in the city and Aqiba rules that any produce rendered unclean in the city must remain. According to this reading, they agree that produce rendered virulently unclean outside the city is removed. This, of course, is the Shammaite position in M. 3:9 D-E. This is further evidence that we are in fact dealing with a second century dispute and not one from the time of the Houses.

[23]By requiring the hide to be buried along with the carcass, M. insures that the farmer will receive no benefit whatsoever from his contaminated deer. The same stringency applies if an animal unfit for the altar is purchased as a peace-offering (M. 1:4). The point is that if the consecrated animal is not fit to be used for the purpose for which it is purchased, no part of the animal may be used for the benefit of the farmer. If the farmer is able to properly use the animal, however, the hide is deemed to be his to do with what he will (M. 1:3).

[24]Nebelah refers to an animal which has died of natural causes, or one which has been slaughtered improperly. This is in distinction to an animal which is suffering from a fatal accident or disease. In this latter case, the animal is in the status of terefah. In either case, the animal is forbidden for consumption. Cf. Lev. 22:18, Deut. 14:21.

[25]Although T. declares that we are talking only about jugs of wine, this is not so specified in M. I have supplied "wine" in the translation simply for convenience's sake and not to indicate agreement with T.'s qualification.

[26]It is not clear from the Hebrew to what the verb "specifying" refers. I have interpreted the passage according to the view of Maim., which is also adopted by TYT and Albeck. Sens. (relying on y. M.S. 3:10) followed by Bert. and TYY adopts an alternative interpretation. According to this second view, "specified" refers to the wine being poured into the jugs and

not to the jugs themselves. Accordingly, the pericope is understood to concern a farmer who is separating second tithe from his wine. If he designates certain wine to be second tithe and then pours it in jugs, the jugs remain unconsecrated. If, however, he pours untithed wine into the jugs and then designates the wine to be second tithe, the jugs also become consecrated, if they were already sealed. I have chosen Maim.'s interpretation over Sens' on the basis of M. 3:13, which clearly relates the law to selling the produce, and not to separating it from untithed produce.

[27]A similar interest as to whether or not a jug becomes consecrated when wine it contains is purchased as second tithe appears in M. 1:3-4. There the law states that in a place in which jugs are normally sold sealed, corked jugs are deemed not to become consecrated when purchased with coins in the status of second tithe. On the other hand, if the jugs are normally sold open, then sealed jugs do become consecrated, whether they are open or sealed. See my comments above. These rulings base the law on local selling customs, a factor not mentioned in the rulings before us. This consideration is introduced in M. 3:13 C-E.

[28]Cf. M. Ter. 4:7. T. Ter. 5:10 carries this principle one step further. It declares that the contents of a jug remain unneutralized only if sealed jugs fall in among sealed jugs, and all the jugs remain sealed. All other combinations result in the consecrated liquid becoming neutralized.

[29]Sealed jugs are one of six (seven) items listed by M. Or. 3:7 which, when consecrated, render consecrated any number of items with which they become mixed.

[30]Cf. T. Ter. 3:8-10.

[31]C-D bring the law here into line with the ruling of M. 1:4M-N.

[32]Lieberman emends our text to put I in the negative, that is, so the jugs will not render consecrated any jugs they are mixed with. Lieberman does not explain why he prefers to emend I and not H, and in fact his emendation seems to run counter to the principle articulated in M. Or. 3:7. Since these jugs are in fact discrete items to which the status of second tithe adheres, it would seem that they should not be neutralized but should render consecrated the entire batch in which they become mixed.

CHAPTER FOUR

[1]A person who is scrupulous about the laws of tithes will separate agricultural gifts even from produce which may already have had tithes removed. In this way, the householder knows that these gifts have been separated at least once. Since what the householder separates may not in fact be in a consecrated status, it is treated more leniently than produce which is certainly in a consecrated status. Joshua's ruling here is consistent with this general principle.

[2]The identities of the foods are from HD.

[3]Delete: "and let him repay" to "what was eaten is eaten" with E. See Lieberman, TK, p. 758.

[4]The consumer returns only the uneaten produce and does not need to replace what he has wrongly eaten. This ruling is in agreement with both Rabbi and Simeon b. Gamaliel in T. 3:10. As we recall, Simeon holds that the consumer must replace produce which he consumed knowing that it was prohibited to him.

[5]For example, a dinar is normally worth twenty-four issars. A moneychanger, however, might buy silver dinars for twenty-three issars and sell silver dinars for twenty-five issars each. The dinars he sells, then, are worth four percent more than the normal value. A this rate, the buyer can purchase six hundred issars of produce for twenty-four dinars (at twenty-five) issars to a dinar) rather than for the usual twenty-five dinars (at twenty-four issars to the dinar).

[6]According to HD, T. is meant to limit the amount of monetary speculation that the farmer can engage in. He takes for granted that if the farmer exchanges his dinars at a more favorable rate than he could in Jerusalem, the profit remains unconsecrated and the farmer may use it to his benefit. Judah's rule would limit the farmer's windfall profit to one roba^{c} per gold dinar. Eliezer allows him to realize a dinar's profit. Cf. Lieberman, TK, p. 752.

[7]The farmer may deconsecrate produce in the status of second tithe by transferring its sanctity to some of his own coins. When he does so, he figures in an additional one-fifth of the produce's value when determining the amount of money he must consecrate. This notion is derived from Lev. 27:31. Cf. M. 4:3.

[8]See Lieberman, TK, p. 753.

[9]As I have noted in my commentary to M. 2:6, coins were deemed to have two values: their face value and the value inherent in their material.

[10]M. 1:2 and T. 1:4 rule that produce in the status of second tithe is not deconsecrated with a slug.

[11]The notion that produce enters the status of heave-offering only when it is actually designated as such is invoked by the Hillelites in T. 2:11.

[12]The language here is awkward. A states that the farmer separates (mpryš) first tithe, leading us to suppose that it is given to the Levites. Yet the end of A indicates that the householder himself eats what should have been given away as first tithe. The point seems to be that the householder must separate produce as first tithe so as to be able to remove heave-offering of the tithe from it. The remainder of the first tithe is now available to be given to any Levite who can demonstrate that first tithe had not yet been taken from the doubtfully tithed produce. If no Levite can establish a claim, the householder is free to consume what he separated as first tithe.

[13]It is assumed that the original owner of produce will always separate heave-offering, even if he is not concerned for the other priestly gifts. The reason is that one who eats heave-offering improperly is subject to death. Since the farmer loses only about two percent of his crop when he separates heave-offering, he does not suffer financially by separating this gift. He is not presumed to have removed first tithe, however, since this comprises ten percent of his crop, and heave-offering of the tithe is therefore also not removed. The householder who acquires doubtfully tithed produce therefore must separate tithe in order that its heave-offering may be removed. He is free to eat the rest of the first tithe, however, since it cannot be shown that he owes it to the Levites.

[14]That this is the expected order for removing priestly gifts is made clear in M. Ter. 3:6 and at T. Ter. 4:10. Both pericopae declare, however, that if these gifts are separated out of order they are still deemed to have been validly separated.

[15]The Hillelites in A held that we can assume that a farmer will always remove second tithe from his produce since fulfilling this law involves no financial loss, the produce remaining in his possession. The householder who acquires doubtfully tithed produce must still designate first tithe and remove heave-offering of the tithes, however. That is, the Hillelites conceive of a situation in which the original farmer separates second tithe but not first tithe. This is exactly the situation sages allow for in F.

[16]The dispute here has a parallel in M. Ter. 3:5, where Simeon and sages have the same opinions as here. See Peck, Terumot.

[17]The law of the added fifth is based on Lev. 27:31, which states, "If a man wishes to redeem any of his tithe, he shall add a fifth to it." Scripture's law leaves it unclear, however, as to whether the farmer owes an extra one-fifth of the original selling price or whether the fee is to be one-fifth of the final price, one-fourth of the original. The issue is disputed in T. M.S. 4:2, with Eleazar taking the former view and Simeon taking the latter.

[18]There are 96 issars in a sela^{c}. If the farmer owes an extra 25% of the selling price of the produce as the added fifth, he must pay one sela^{c} and 24 issars. If he owes 20% as the added fifth, he pays only one sela^{c} and approximately 19

issars. I have assumed the latter figure is in the commentary since it is the least the farmer will have to pay. It thus is the minimal difference between what the farmer will pay and what others must pay.

[19]Canaanite servants, that is, non-Hebrew slaves, are mentioned only six other times in M. They are acquired by purchase, deed or through serving their owner (M. Qid. 1:3). If someone injures a Canaanite slave, the owner is owed compensation (M. B.Q. 8:3), unless the owner himself caused the injury (M. B.Q. 8:5). The owner also has a right to dedicate them to the priests (ḥrm) just as he may do with his cattle (M. Arak. 8:4). Whatever a Canaanite servant finds belongs to his owner (M. B.M. 1:5) and he is always deemed to be acting only on behalf of his owner (M. Er. 7:6). It is clear, then, that the law in M. regards the Canaanite servant as the personal property of his owner, without individual status. A Hebrew servant, in contrast, is deemed to be an autonomous agent who is legally responsible for his own acts.

[19a]Cf. T. 3:15. We assume that the farmer, even if he is an ᶜam ha'areṣ, always will remove heave-offering from produce he intends to sell. We do not assume he will separate heave-offering of the tithe, however, since he does not separate first tithe. The consumer, therefore, must separate heave-offering of the tithe so that all heave-offering in the produce is accounted for.

[20]The concept of poorman's tithe is derived from Deut. 14:28-29. It is separated in the third and sixth years of the Sabbatical cycle in place of second tithe and is set aside for the poor. This produce is not given to the farmer's dependents since they must be supported out of the farmer's own produce.

[21]According to Lieberman, the purchaser may separate heave-offering of the tithe from his produce for other untithed produce he has. He will thus be wrongly separating tithes from tithed produce for untithed produce.

[22]HD interprets the pericope differently. A passerby, seeing a ḥaber separating tithes from the produce of an ᶜam ha'areṣ, might wrongly assume that one may separate tithes from doubtfully tithed produce for certainly tithed produce. To prevent giving this wrong impression, a ḥaber should separate tithes only from produce the ᶜam ha'areṣ brings out of his house and which he affirms is ṭebel, that is, certainly untithed. Similarly, a ḥaber should not transfer the status of second tithe to coins supplied by an ᶜam ha'areṣ lest a passerby assume that one can always trust the coins of an ᶜam ha'areṣ to be unconsecrated.

[23]Lieberman suggests emending qtn (minor) to śṭn or śytwn (wholesale merchant).

[24]This ruling runs counter to the concern of T. 4:3. There the person recruited to act as the buyer had to have full control over the coins before the sale was deemed to be valid.

[25]HD declares that this works only if the farmer originally intended to give the coin to the merchant so that he could buy the farmer's produce. It is the farmer's intention to sell the

produce to someone else that exempts him from paying the added fifth. The act of finally giving the coin to the merchant has no legal value, but is simply the final acting out of the farmer's intention.

[26]The interpretation here follows Bert., TYT, Sens, TYY and Albeck. Maimonides, however, reads the Hebrew to mean that the farmer is purchasing produce for use as second tithe. In the first case, according to this view, the farmer selects a sela^{c}'s worth of unconsecrated produce, which subsequently doubles in value. The farmer gives to the merchant only one sela^{c} for the produce, but has, in fact, deconsecrated two sela^{c}s' worth of coins. He thus has an extra sela^{c} in his possession which he may spend however he wishes. In D-F the farmer selects two sela^{c}s' worth of produce, which subsequently become cheaper. The produce, then, effects the deconsecration of only one sela^{c} in coin. The buyer, however, still must pay two sela^{c}s. This leaves an uncertainty as to which of the coins is in fact deconsecrated. To avoid confusion, the farmer pays one sela^{c} in consecrated coin, which becomes deconsecrated, and one sela^{c} in unconsecrated coin. The seller is free to use both coins for secular purposes.

[27]M. Qid. 1:1 declares that the bride-price can take one of three forms: an object of value (ksp), a marriage certificate, or sexual intercourse.

[28]A similar dispute appears in T. Qid. 2:8. The householder simply hands over the bride-price without any mention of marriage. As we would expect, Yosé deems the woman to be married, and Judah deems her not to be married. Rabbi declares that if they were already talking of marriage when he handed over the bride-price, she is married even without the householder making a formal declaration. This is in line with Yosé's thinking in our pericope.

[29]The text of G does not explicitly differentiate between doubtful second tithe and certain second tithe. That such a distinction is in fact intended is clear from the language of H, "in all cases," and the explicit reference made in I.

[30]The addition of the words "in Jerusalem" makes good sense. Under the Mishnaic system, a large number of coins in the status of second tithe would be taken to Jerusalem. There would be a good chance, therefore, that coins found in that city are indeed in a consecrated status.

[31]The abbreviations are resolved in y. M.S. 4:9 as follows: "Aleph" stands for what is removed first; "dalet" stands for dema'i (Lieberman suggests instead dema^{c}); "resh" for what is removed first (r'šwn); "taw" for Terumah. "Yod" indicates a tenth, that is a tithe, and "mem" abbreviates m^{c}śr. Lieberman, TK, p. 776, claims that "pe" stands for pidyon (that which is to be redeemed as second tithe) and "shin" means sheni, that is, second [tithe].

[32]T. 3:17 has the complementary case. Simeon and sages disagree as to whether or not a farmer who is designating second tithe in a batch of produce must specify a particular portion of the batch. Simeon holds that the farmer does not need to

designate a particular part of the batch as second tithe. Sages here would agree. In contrast, sages in T. 3:17 require a specific portion of produce to be designated, a view congruent to Yosé here. Cf. M. Ter. 3:5.

[33]The view here is congruent to the view of M. 4:10E-F (T. 5:1A-F) concerning inscriptions. There we saw that produce is deemed consecrated if its container bears even an ambiguous label. Likewise here the produce is to be considered consecrated even if there is doubt that the farmer's statement applies to the coins. The opposing view in F is then paralleled by that of sages in T. 5:1G-K. F claims that the statement is considered to be probative only if it clearly refers to found coins. Sages in T. 5:1G-K claim that a label is probative of the status of produce only if it clearly refers to a consecrated status.

[34]The reasoning here is similar to that of Yosé in M. 4:10-11. He holds there that if there is some doubt about an inscription, the produce it is associated with is deemed unconsecrated.

[35]The meaning of this rule is unclear because of the elliptical nature of the language. The parallel in B. San. 30a offers a different wording. There this rule reads as follows:

> [Concerning] what is accessible to him (kl šbydw lyṭln) — his words have effect.
> [Concerning] what is not accessible to him — his words have no effect.

This wording makes the same point as T.'s, but in a clearer way. If the witness wanted the coins for his own use, he could simply take them. By disclosing their location, he indicates that he has no ulterior motive. We thus believe him when he says they are consecrated.

GRA proposes that we reverse the apodoses of the general rule so that it reads as follows:

M. [Concerning anything they can find [independently] — his statement has effect.
N. [Concerning] anything they cannot verify [independently] — his statement has no effect.

This reading brings L-N into line with H-K. According to this reading, the point is that if the witness gives specific information, so that it is possible to check on his statement, we take his testimony seriously. On the other hand, if he gives only a general location, which we cannot verify, we disregard his claim.

Lieberman interprets the rule to mean that a single witness has the power to prohibit use of an object only if that object is not in the hands of its owner. That is, if the sons have access to the father's coins, the testimony of one witness that the coins are consecrated has no effect. Cf. TK, p. 782.

[36]H-K in fact can be understood as an autonomous unit independent of L-N. Read in this way, the point is that a statement must refer to specific coins for those coins to be deemed consecrated. At H-I the speaker does not name a specific location for the coins he claims are consecrated. As we saw in T. 5:5F, such a statement is of no value. The statement in J-K,

on the other hand, is specific. Coins found in that location are deemed to be consecrated, in line with the ruling of M. 4:12D-I. By appending the rule of L-N to this unit, the redactor clearly wants us to read this material as making the same point as O-S.

CHAPTER FIVE

[1]TYY relates each of the devices used for marking the area with the type of prohibition it signals. Clods of earth support growth only when properly prepared. So too grapes of a vineyard's fourth year are edible when properly prepared, by being deconsecrated with coins. Clay will not support growth and thus is used to mark grapes of a vineyard's first three years, since such grapes are not to be used at all. Lime marks off a graveyard because it is symbolic of bones.

[2]The notion that the fruit a tree bears during its first three years of maturity is forbidden is Scriptural. Lev. 19:23-24 state: "When you come into the land and plant any fruit trees you must consider forbidden (w^{c}rltm ^{c}rltw) its fruit; they shall be improper for you three years and are not to be eaten."

[3]See Nu. 19:16. M. makes several allusions to the practice of regularly marking off grave areas. See, for example, M. Sheq. 1:1, M. M.Q. 1:2.

[4]M. thus assumes that even though the produce is deemed ownerless, the owner of the vineyard is responsible for bringing its fourth year crop to Jerusalem for consumption. The issue of whether or not the farmer is in fact the owner of this produce at all is taken up in M. 5:3.

[5]The consecrated status of the grapes is transferred only after the grapes have been plucked. This is so because the produce itself only takes on the status of consecration when it is harvested; see M. 5:5. This is also the case with produce to be designated as second tithe (cf. M. 1:5).

[6]See y. M.S. 5:1.

[7]Lieberman, TK, pp. 782-783, holds that the danger is that someone digging in the area will uncover blood and will be thereby rendered unclean. Blood is poured on the site to make the danger apparent. When the blood poured on the ground is fully absorbed, we can assume the same is true of the corpse's blood, and the area is deemed safe for excavation (cf. M. Oh. 3:2).

[8]Lieberman, TK, p. 703.

[9]Jastrow, p. 51, claims that the name should read Ayyelet. I have been unable to find any other reference to such a place, however. Neubauer, La Géographie du Talmud (Paris: 1868), pp. 119-120, considers a number of possible locations for the Elat mentioned here, including Elusa in Trans-Jordan, the

Hellenistic city of Eleuthéropolis, and the Elah Valley, all of which he rejects. I can offer no new suggestion as to the site referred to in our pericope.

[10]Lieberman holds that Simeon (J-K) does not want the farmer to give fourth year produce to friends and relatives but rather wants to encourage the use of the grapes to decorate the market. According to Lieberman, Simeon reasons that giving the produce away will leave none over for decoration. The farmer therefore must sell the consecrated produce, thereby making it available for use as decoration. Lieberman goes on to make the claim that what is purchased with the coins the farmer receives takes on the status of second tithe. According to this view, then, the status of fourth year growth and the status of second tithe are in some sense interchangeable. M.-T. hardly seem to bear this out, however.

[11]The issue of whether or not grapes of a vine's fourth year must be processed into wine before being taken to Jerusalem is the subject of a Houses' dispute in T. 5:10. The point of the Hillelite position is that the farmer must increase the value of his consecrated food as much as possible. In the case of grapes, this means processing them into wine.

[12]According to y. Peah 7:6, the point of Rabbi's and Simeon's disagreement is whether or not the Shammaites derive the laws governing fourth year produce from the laws of second tithe. Rabbi presumes that the Shammaites do draw an analogy between fourth year produce and second tithe. During the Sabbatical year, second tithe is not separated, so that the laws of removal and the added fifth do not apply to it. Similarly, say the Shammaites, such laws do not apply to fourth year produce in the Sabbatical year. Simeon holds that the Shammaite view is that fourth year produce is never like second tithe, and so the laws of removal and added fifth never apply to it.

[13]Although M. employs language which seems to distinguish between grapes of a vine's fourth year and fruit of the fourth year of other plantings, these phrases in fact refer to the same consecrated status. The notion that the produce of any planting's fourth year is consecrated is derived from Lev. 19:23-25, which states, "When you enter the land and plant any fruit tree, consider its fruit taboo (lit., uncircumcised, w^c^rltm ^c^rltw 't pryw), let it be for you taboo for three years and be uneaten. And in the fourth year all of its produce is holy, dedicated (hlwlym) to the Lord." M.'s discussion of such produce, almost all of which is contained in our tractate, does not indicate that the law treating grapes is different from that treating other produce in this regard.

[14]Produce of a planting's fourth year is unlike produce in the status of second tithe in that it may be sold before it is harvested. This is so because produce in the status of second tithe takes on a consecrate status only after it is picked (M. 1:5). T., as we shall see, disagrees, holding that in this regard, produce of a planting's fourth year is like produce in the status of second tithe and should be sold only after it is harvested.

[15]During the Sabbatical year, the farmer is allowed to make use of produce which grows in his field without being purposely

cultivated. Of this produce, however, he may harvest at one time only as much as he intends to eat at one sitting. Thus, while agricultural labor is generally prohibited during this year, M. does accord the owner some measure of ownership over produce which grows in his field during this year. He may sell such produce, for example, transferring its special status to coins (M. Sheb. 7:1-3) and he is responsible for seeing to it that produce in its fourth year of growth is not improperly picked from his field and eaten (M. M.S. 5:1).

[16]Maim., Bert., TYT, MS, and Albeck all read E to mean that the seller of ownerless produce sells it at its full market price minus the cost of harvesting. According to this view, ownerless property is unlike seventh year produce in that one who takes it over may harvest it, but it is unlike owned produce in that one who purchases it has no right to guard it or perform other acts of labor on its behalf. I do not see why this should be the case, since the purchaser of an ownerless field should certainly be able to improve or guard his investment. Furthermore, I think this reading of the pericope misinterprets the language of E, which seems to imply that the seller only has a right to limited compensation (<u>śkr</u>) not a right to its full market value.

[17]Grapevines, in order to reach full productivity, must be pruned back annually until they reach maturity. This allows for maximum growth and development of a few select shoots rather than leaving the vine to produce a tangle of smaller shoots. A full treatment of this subject is founde in SCH V, pp. 1384-1395. See also Columella, <u>On Agriculture</u>, Bk. IV, Chap. 22-28.

MB interprets <u>gmm</u> to mean "uproot." According to this reading, the pericope concerns a farmer who uproots an entire vine and wishes to sell the consecrated grapes so as to receive coins he can take to Jerusalem. The question for MB is whether or not such an uprooted vine is considered attached to the ground because of the clods of earth still clinging to it. If so, the grapes are deemed unharvested and so may not be sold (T. 5:19).

[18]Lieberman claims that the discussion here concerns vines which, in their fourth year of growth, have not produced a crop. Citing B.R. 38:9, he states that such vines were routinely pruned back in order to increase future yields. This practice is prohibited, according to Lieberman, lest the farmer become accustomed to pruning his vines in their fourth year of growth and mistakenly cuts back growing grapes, thus transgressing the law of fourth year produce.

[19]Cf. M. 5:3F, where the Hillelites declare that all grapes of a vine's fourth year must be taken to the winepress.

[20]There have been attempts to harmonize this ruling with M. Several authorities claim that one normally may not sell fourth year produce that is still attached to the ground, in accordance with T. 5:19C. This is so because while it is still attached to the ground, its value cannot be determined. If a selling price can be set, however, and the produce is sold, the sale is legal, in accordance with M. 5:4-5. See Lieberman, <u>TK</u>, p. 786.

[21]Lieberman holds that the problem here is that produce grown during the Sabbatical year must be removed from the householder's domain before Passover of that year (M. 5:6). This requirement diminishes the amount of time available to the farmer properly to eat in Jerusalem, or redeem, consecrated produce grown during this year. The Shammaites thus declare that vines should be planted so that their fourth annual yield will not coincide with the Sabbatical year. Lieberman does not explain why this same consideration does not apply to a crop planted in the first year of the Sabbatical cycle, which also will reach its fourth year of growth during a year in which the law of removal takes effect.

[22]There is considerable disagreement in mss. as to whether the proper reading is r'šwn or 'ḥrwn. Sacks-Hutner supply an extensive overview of the various manuscript traditions, pp. 302-303. I have chosen 'ḥrwn since this is the reading found in most manuscripts of M. TYT argues that this time is more advantageous to the farmer than would be the first day of the holiday since it allows the farmer time to distribute his agricultural gifts while he is in Jerusalem for the festival. Only when the festival is drawing to a close must the farmer destroy whatever agricultural gifts he has been unable to distribute.

[23]Rice and various herbs and spices were planted in early summer and were harvested only in the fall of the following year, five months later. Felix discusses a variety of plants which he holds were planted in the summer and harvested in the fall, p. 105f.

[24]My explanation of why Passover of the fourth and seventh years is chosen follows that of y. M.S. 5:5 and is found also in Maim., Sens, Bert. and TYT.

[25]Accordingly, the requirement to remove such produce from one's domain does not fall on the priests, Levites or the poor who receive such gifts (MR). The point, as we stated, is solely to limit the farmer's keeping to himself consecrated foods, which belong to others.

[26]M. Bik. 2:2 also rules that in matters of removal, first fruits is like produce in the status of second tithe.

[27]My interpretation follows Bert., TYY and Albeck. Maim. and MR explain the Hillelite opinion on the grounds that the cooked dish will soon go bad anyway if not eaten immediately. There is no need, therefore, for the householder to make a special effort to throw it out. This view is based on y. M.S. 5:6, which distinguishes pastry and oil, which must be removed, and spiced dishes or wine, which may remain. The point, according to MR, is that the latter will soon spoil and thus will, in effect, be destroyed. This second view is questionable on two grounds. First, it is based on a distinction unknown to M. Second, it ignores the fact that the householder is obligated to declare in his confession that all consecrated produce has been removed from his domain. It would be a violation of the oath to leave some consecrated food in the house on the grounds that it will soon go bad.

[28]If produce in the status of second tithe is cooked along with unconsecrated produce, for example, the completed dish is deemed to have in it food of both types. If it is sold, an

appropriate amount of money received is in the status of second tithe, and the rest is unconsecrated, cf. M. 2:1. Similarly, if produce in the status of heave-offering is cooked in a dish of unconsecrated food, the entire dish becomes forbidden to non-priests if the consecrated produce imparts its flavor to the whole (M. Ter. 10:1-5). The only exception is if the consecrated ingredient is of such minute proportion (one part in a hundred) that it itself is neutralized, losing its consecrated status (M. Ter. 4:7).

[29]Cf. T. R.H. 1:8, T. Sheb. 2:3. Lieberman offers a slightly different calculation. He claims that the tree is considered to enter its second year of growth on the fifteenth of Shevat of the Sabbatical year, a year earlier than we have claimed. As a consequence, the tree becomes subject to the removal of tithes on the fifteenth of Shevat of year three of the Sabbatical cycle. Second cycle is not separated this year, however, since poorman's tithe is owed instead. Second tithe is first removed from the tree's fruit during the fourth year of the Sabbatical cycle and is therefore not subject to removal until the next Sabbatical year. HD seems to hold that the tree's age depends on calendrical years. We do not, in his view, count the growth of the tree at all during the Sabbatical year. The tree is thus in the status of ᶜorlah through year three of the Sabbatical cycle and is consecrated as well for the fourth year of the cycle, as fourth year produce. Second tithe, in his view, is first removed during the fifth year of the Sabbatical cycle and, consequently, is not subject to removal until the Sabbatical year.

[30]The Houses here are depicted as active after the destruction of the Temple in 70. This has occurred once before in our tractate, T. 3:13-14. In both cases, the issues about which the Houses debate are not taken up again in our tractate. We therefore cannot apply the normal criteria for testing these attributions.

[31]Albeck relates the Hillelite opinion here to the law of M. 1:5E and 1:6G, which rule that after the Temple's destruction, produce purposely purchased as second tithe outside of Jerusalem must be left to rot. The point, as Albeck sees it, is that in both instances, consecrated produce is not sold but is itself destroyed. These laws, however, are not congruent. Produce purchased as second tithe, such as that of M. 1:5-6, may not itself be deconsecrated (M. 3:10) under any circumstances. There is, consequently, no question that the farmer must allow this produce to rot if he cannot eat it in Jerusalem. The present pericope, however, deals with produce originally designated as second tithe. Under normal circumstances, such produce may be sold. It is only about originally designated produce, then, that the question of whether or not to sell can arise.

[32]Rabbinic documents preserve two such letters allegedly from the time before the destruction of the Temple or shortly thereafter. The first letter appears in T. Sanh. 21b and parallels (y. M.S. 5:4; y. Sanh. 1:2; b. Sanh. 11b). This letter, as it appears in T. Sanh., reads as follows:

A. Mᶜśh b: Rabban Gamaliel and the Elders who were sitting on the steps of the Temple mount and Yoḥanan, the scribe, was before them.

B. He said to him [i.e., to Yoḥanan], "Write to our brethren of Upper Galilee and Lower Galilee, 'May our peace increase. We [thereby] inform you that the time of removal has arrived, [the time] to take out tithes from the olive-vat.'

C. "And [write] to our brethren of the Upper South and the Lower South, 'May your peace increase. We [hereby] inform you that the time of removal has arrived, [the time] to take out tithes from the sheaves of wheat.'

D. "And [write] to our brethren of the diaspora in Babylonia and the diaspora in Medea and all the other diasporas of Israel, 'May your peace increase. We [hereby] inform you that [since] the pigeons are [still] tender and lambs are [still] weak and [since] spring has not yet arrived, it seems to me and my colleagues that we should add thirty days to this year [i.e., declare a leap-year].'"

From this letter it is clear that farmers were expected to separate agricultural gifts before the time of removal from produce which was already harvested and thus liable for the separation of these gifts. Both Judah and Aqiba would agree with the author of this letter.

The second letter, found in Mid. Tan. to Deut. 26:13, is ascribed to Rabban Simeon b. Gamaliel and Yoḥanan ben Zakkai. Its relevant part reads as follows:

> Let it be known to you that the fourth year has arrived and as of yet [produce] dedicated to heaven (<u>qdšy šmym</u>) has not been destroyed. Rather (<u>'l'š-</u>) you should hurry to bring the five sheaves which are required (<u>šhn mᶜkbyn</u>) for the confession. . . .

I have included this letter for the sake of completeness, since it does refer to the time of removal. It is not clear to me why five sheaves are to be brought to Jerusalem at this time.

[33] MR holds that in A's view plants still growing in the field at the time of removal constitute a special case. Agricultural gifts need not be separated from such plants before the current time of removal. If they are separated after the time of removal, however, they must be immediately disposed of and may not be kept until the next time of removal. To save the farmer the inconvenience of having quickly to distribute heave-offering and first tithe and of having immediately to eat second tithe in Jerusalem, A declares that these gifts should be designated before the time of removal arrives.

[34] M. Ma. Chapter One suggests two different times in their growth and processing at which produce becomes liable to the removal of heave-offering and tithes. The first half of the chapter (M. 1:2-5) suggests that agricultural gifts are separated from the fruit only after it is mature enough to eat. This is the view ascribed here to Aqiba. The second half of the chapter claims that heave-offerings and tithes are removed when the produce has been harvested and is in fact ready to be eaten. In either case, the point is that agricultural gifts, which are to be eaten by their recipients, are to be separated only from produce which is edible. For a more detailed discussion, see Jaffee, <u>Maaserot</u>.

[35]Maimonides and MS claim that the pericope contains two distinct laws. According to this view, A-B declare that a farmer who cannot separate heave-offering and tithes from his produce must at least designate these gifts in his produce. C-E then allows the farmer to transfer his agricultural gifts to their proper recipients as well, if possible. Bert. and TYY offer a somewhat different interpretation. They claim that in A-B the priestly gifts have already been separated. The oral declaration referred to in B designates only the specific recipients of the already consecrated food, as is illustrated in C-E.

[36]Meir's claim is backed by several verses of Scripture, including Nu. 18:20-24; Deut. 10:9; 12:12; Jos. 13:14, 23. The whole argument is unusual since the priests are manifestly not farmers and thus do not separate agricultural gifts to begin with. The Levites do receive first tithe from which they must separate heave-offering of the tithe, so they do come under the scope of the law.

[37]The Hebrew here is obscure. The translation I have supplied, which represents also the understanding of T., is faithful to the Hebrew but does not make sense in its Scriptural context. The Septuagint reads instead, "You have sated the missiles of your quiver." RSV has, "and put the arrows to the string. Selah."

[38]The term, dema'i, refers to produce purchased from one who is not trusted properly to remove agricultural gifts and about which there is thus doubt as to whether heave-offering and tithes have been separated. It seems inappropriate to translate the term as "doubtfully tithed produce" here because the point of the pericope is that there is no longer doubt about which gifts have been separated from the food. I therefore have taken the word to refer to food purchased from one who is not known to remove all agricultural gifts from what he sells.

[39]Neusner has argued that in Tannaitic times Yoḥanan was held in high regard and was deemed to provide valid legal precedent. Rabbinic attitudes towards this figure changed only in Amoraic times, when he came to be pictured as a Sadducee. See Pharisees I, pp. 160-176.

[40]Lieberman, in Hellenism in Jewish Palestine, pp. 139-143, discusses this series of reforms and the reasons T. gives for their enactment. His basic thesis is that T. often ignores important aspects of the reforms and that therefore its explanations of them is inadequate. Lieberman then adduces his own account of why these reforms were enacted, drawing upon statements in rabbinic literature and upon other sources pertinent to the historical period in question. His argument fails, however, since it depends on his unwarranted assertion that T. preserves only Yoḥanan's public statements and not his true reasoning. Furthermore, Lieberman's uncritical use of later rabbinic materials makes his own explanations appear to be arbitrary. Since a fuller treatment of Lieberman's essay would involve us in a discussion of problems in Talmudic historiography, it need not detain us here.

INDEXES

PERICOPAE

Mishnah

Tosefta

Tosefta (Continued)

NAMES AND TOPICS